FREE HEALTH
Your Own "NHS" Natural Health Service

Michael Lingard

"Is there really a pill for every ill or can we all learn how to live a healthy life with the right guidance and support?"

"The doctor of the future will give no medicine, but will interest his patients in the care of the human frame, in diet, and in the cause and prevention of disease.
-Thomas A. Edison. 1847-1931

"Nothing will benefit human health and increase the chances for survival of life on Earth as much as the evolution to a vegetarian diet."
- Albert Einstein. 1879-1955

" Example is not the main thing in influencing others; it's the only thing."
- Albert Schweitzer 1875-1965

" It has become appallingly obvious that our technology has exceeded our humanity." - Albert Einstein 1847-1931

"The most powerful way to restore health is often to do nothing- intelligently."
- Dr Alec Burton 1930-2016

"There is a Ministry of Health, and the Department of Health. Surely, in the long view, this should be The Ministry, and Department of Public Happiness?"
- Lord Horder 1871-1955

"One of the essential qualities of the clinician is interest in humanity, for the secret of the care of the patient, is in caring for the patient"
- Dr Francis W Peabody 1881-1927

I am just one among the many thousands of individuals, either medically trained or general members of the public interested in enjoying a long, active and healthy life who are trying to take more responsibility for their own health and wellbeing. This small book is my contribution based on a distillation of the essence of many great minds who have studied health and its origins in the past.

Published by Lulu .com in 2020

ISBN 978-0-244-54999-2

Text copyright © 2020, Michael Lingard

CONTENTS

Forward

This small book will give you many ideas that you might want to take on board to improve your health and the health of others. I hope it will help you understand that the greatest secret kept from us by the medical establishment is that good health is the norm if we help ourselves, how else could the human race have survived thousands of years without the medical care we all assume is essential for our health and survival?

It is a record of over forty years the work of a passionate health provider searching for "the magic bullet" to give us all vibrant health and a long active life. As you read the book you will realize there is no "magic bullet", which may be a disappointment, but there are a few simple lifestyle changes that will give most of us those healthful outcomes.

The grand question, the "elephant in the room", is, are you ready and willing to make changes in your lifestyle? If you are, even in small stages, over a number of years, you can expect better health, lower risk of developing any serious chronic diseases, more energy, a calm mind and a long active life.

The decision will be yours and yours alone. No doctor or health Guru, however brilliant, can make you healthy. This is an individual's challenge and responsibility, others can only support and guide you to the best of their ability. It is hoped that this small book will provide some of that guidance and direction that will benefit you on your journey to vibrant health and a long active life.

Read on and enjoy.

Chapter 1

Towards A Better & Healthier Nation

We are at a tipping point, a crisis stage in our National Health Service. Everyone agrees it is no longer able to continue in its present form, this is not news as, over a decade ago when the King's Fund gave a presentation at the International Health Workers Conference in Brighton, they had been given a remit from the Government to find ways to change the NHS from a "sickness service" to a "health service". The essence of their presentation was that this task was like turning a giant tanker around; not impossible but it takes a great effort and a long time. Fifteen years on and little has changed, except the situation is even more desperate now.

If we believe the NHS will not radically change in the future, what can we do ourselves to improve our own and the nation's health? There is a growing interest amongst many doctors for encouraging lifestyle changes across the population to promote better health, it is called "social prescribing" but whatever the name given, lifestyle changes are fortunately outside the normal remit of our NHS.

What are these lifestyle changes? They may include: basic health promoting changes involving amongst other things, what we eat, how we exercise, how we breathe, how we relax and how we enjoy our lives as fully as possible. This is the field of expertise and experience of all those in the natural therapy world who have worked in this way over the past many decades. These therapists offer a vast potential for changing the healthcare provision that is so urgently needed. All physical therapists including chiropractors, physiotherapists, sports massage therapists, Bowen Technique therapists, occupational therapists, cranio-sacral therapists and many more can develop and provide this service if they are given the right support by the public and government.

Some physical therapists are content with the service they offer and may accept all the constraints imposed on their practice of healthcare that their governing bodies impose upon them. There are those others who want to provide a far more holistic support for their patients, and it is this group who would benefit from joining forces with other like-minded professionals to create a significant force for change. There is such an organization that seeks to provide just this form of association of the tens of thousands of physical therapists. It is called the Democratic Orthopathic Council or DOC[1]. Only time will tell whether there is sufficient support for such a force for change, meanwhile perhaps the best advice one can give to everyone is learn how to be more responsible for your own health. If enough of us learn to make the necessary lifestyle changes, not only will we have a healthier nation, with less pressure on our NHS but also we shall be helping the survival of the planet.

Now that's worth a little effort for future generations!

Chapter 2

The Real Health Assessment

Most health checks focus on pathology, screening for diseases and markers of possible future health problems, and little information is offered the client on health promotion or possible therapies that would help them towards better health and wellbeing.

As an Orthopath, I am always trying to improve the health of my patients and in particular teaching them how to help them help themselves to better health.

A Real Health Assessment should be designed to identify changes of lifestyle to improve health and fitness, and to reduce the chance of serious disease in future. Doctors are not trained in health promotion, most of their training being devoted to pathology and medical treatment with drugs or surgery. It is this neglected area of healthcare that a Real Health Assessment is about. Far more than simply identification and prevention of disease or sickness, it should seek to guide individuals to the health and zest that most of us only experienced when very young.

There is a widespread view that we should all feel less healthy as we get older, this is not necessarily true, some of my "youngest patients" are in their 80s. So what's so different? I check their mechanics, the body connection, this is the normal field of therapy for physical therapists and seriously neglected in our NHS. I check their diet, the food connection, "we are what we eat". I only added this modality to my practice after discovering the work of the China Study. I use a simple twelve question survey to screen every patient, this takes less than five minutes, after which I give an A4 sheet of useful notes as a guide to better eating and better health. I check their breathing, the breath connection. Breathing is the most important activity in our lives; we can survive the three weeks without eating, three days without drinking, but only three minutes without breathing! Also it would appear that all mammals, that include you and me, have a given number of breaths in a lifetime, so it matters that we use them well! Read more about this in chapter 6, Half a Trillion Breaths!

I trained as a Buteyko Breath Educator fifteen years ago and I now check every patient's breathing quality as part of an ongoing research programme and give them a simple guide to better breathing and better health, if they don't want to have a full training course.

This research programme has identified a relatively unrecognized relationship between "what we eat" and "how we breathe". Did you know that what you eat will change the way you breathe, and how you breathe will change the way you choose your food to eat? This link I discovered after gathering a sample of over two hundred patients over the past three years. It came as a surprise to me even though I've been working in the field of breath training for over fifteen years, and as a whole plant nutritionist for the last three years.

Since a poor diet can lead to most modern diseases, from allergies and diabetes to cancers, and poor breathing is a major factor in the development of asthma, hypertension, panic attacks, ME, heart problems, gut problems and many more conditions, even orthodontic problems[2], this link between the two needs far more attention.

To explain the physiological connection between poor diet and poor breathing is fairly easy; a diet comprised of too much meat, dairy foods, soft drinks and highly processed foods often leads to metabolic acidosis, the body becomes too acid. To correct this the body can either draw calcium from your bones or cause you to hyperventilate and get rid of carbon dioxide to neutralize the acidity. Thus it is important, when teaching people to breathe normally and to stop hyperventilating, to ensure they are eating a good diet. The connection between their breathing and diet is more subtle, when we breathe normally the body is optimally oxygenated and we are usually calm and free of stress. This appears to make us more thoughtful of what we eat, we tend not to snack on junk foods.

Chapter 3

Real Health Insurance for Free!

Millions of people spend millions of £'s on health insurance, but it doesn't guarantee health, reduce the chance of sickness or improve the chance of better health. Most health insurance systems don't even pay for your medical treatment if you suffer from a health problem you have had before, which is the most likely thing to happen. It is rare for any Health Insurance scheme to pay towards your health promotion and education, their very business depends on you being sick!

There is always the chance that even with the very best effort on the individual's part to stay healthy, they may still fall ill by chance or because of factors outside of their control, this is where our NHS should be able to offer the very best service.

Unfortunately because of the mounting demands on the NHS, largely arising from a lack of health promotion and education, it is becoming increasingly difficult to provide the quality of service that most doctors and health professionals would like to give to all their patients.

Those people with their own private medical insurance may not suffer this way but they may be encouraged to accept more medical intervention than possibly needed and often if there is a major problem requiring complex, costly intervention, they will find themselves back with our NHS.

There is another Real Health Insurance that does reduce the chance of serious chronic diseases, increases the chance of a long healthy, active life and most important, it doesn't cost millions and can be free.

You can't buy it off the shelf, you don't need to take pills and potions, it is based on sound research and good science and with just a little support you can provide it all yourself!

Now is that too good to be true?

Well no, but you will need to make some effort yourself to change your lifestyle. Only you can decide whether that's a "price" worth paying.

There is a well-kept secret in medicine; that good health is the norm and that sickness should be the exception.

However there is a vast ever-growing industry based on disease and sickness that would not survive if everyone took more care of them-selves and their families.

The Healthcare profession is a little different from other professions. The measure of the success of any practitioner in the healthcare business is how well their clients can manage without their services! Not always the best business concept if profit is the bottom line! This may have been the reason why the Chinese doctors used only to be paid while their patients were well and would treat for free if they became ill! Such a system would spell financial disaster for the medical establishment today.

The best healthcare practitioner teaches people how to care for their own health and how to rely less on others for support or to need pills, potions and drugs to stay well. When my patients visit me only once a year or even once every few years I feel I have done a good job of helping them to help themselves. This book is based on a free podcast that I decided would benefit all my patients, called "Your Health in Your Hands" this free health podcast show that brings you health promoting information, tips and ideas that will change your life and help you take more responsibility for your health. It is a ongoing project for me and coming episodes will introduce people to many health-promoting ideas, exercises, videos, and to other leaders in this field who are passionate about teaching people how to help themselves to stay healthy.

It is interesting to remember the etymology or original meaning of Doctor, very simply, teacher! When there were far fewer possible drugs and medical interventions available the best physicians were great health educators and supporters.

Chapter 4

Health is Based on Family & Community

As part of my passion to discover the fundamentals of health and wellbeing I studied the work of a remarkable doctor, Dr Scott-Williamson. In his later years as a brilliant pathologist he asked himself a very simple question that would become the basis for his work and research for the rest of his life. As a pathologist he was reasonably happy that he knew the aetiology of most common diseases, what were the key causes of each disease, that was his profession, but when he asked himself what was the aetiology of health, what in other words were the key causes of health, he had no answer.

He then began a search through all published research papers in this country and abroad but could find little on this subject. What he knew, as most of us would agree, is that health isn't just what is left after we remove all diseases, it is far more than that.

To summarize over twenty years research, in just a few words, what he discovered was that health was based on a child growing up with a father and mother in a community that supports them as a family. He also took for granted that the child needed an adequate diet, exercise and an environment where it could develop its own talents as fully as possible.

This conclusion was based on the experience of over a thousand families who were part of what became called the Peckham Experiment[7]. In Peckham, in London, a large modern building was designed with swimming pool, gymnasium, cafeteria, crèche and areas for outdoor recreations as well as a medical unit where all participating families has their annual health audit. Anyone who was found to need medical attention or tests was advised to see their own doctor, whether they did or not was their decision. This was all part of encouraging self-responsibility for their own health and that of their family.

By every medical measure, the health of all participants improved year on year. It was regarded as the most important medical research of its time and the centre had thousands of visitors from home and abroad every year, even including the Queen.

When the NHS was being planned by Aneurin Bevin, a group approached the Government to propose that similar centres should be established throughout the country to improve the health and wellbeing of the nation. The other group that approached the government suggested that if people had better access to modern medicines this would ensure a healthier nation. We now know who won the day and only now are we seeing a growing recognition that drugs and medical interventions do not always lead to improved health and cure of the disease but usually they simply help manage disease and ameliorate symptoms.

If Dr Scott-Williamson was right, that our own health and the health of our nation is built on family and community, it may explain why there is so much sickness in the UK. As a nation, we may lead the world in this, as we have a record of ever increasing numbers of broken families and poor community support; those two prime factors that Dr Scott-Williamson and his co-worker Dr Innes-Pearce proved were essential for the development of a healthy child. We are seeing an worrying rise in childhood mental problems through stress and anxiety, besides the adverse effects of diet and lack of exercise leading to an epidemic of obesity and related diseases.

The very word health is derived from Hale, old English for whole. The rationale behind the Peckham Experiment findings is that for a child to fully develop as a whole person they need a secure environment of a mother and father, as only when we have the union of a male and female is there a whole human. Every man is essentially half a human, as is every woman, but together they represent a model of a whole human that the child learns from.

The rich environment of many families meeting, working and playing together in the Peckham Centre increased the possibility of children growing to their fullest potential.

All the latest research is pointing to the importance of our childhood development from birth to early schooling, how this experience can strengthen or weaken a child for life. As another pioneer in this area Dr Glen Doman[8] says, "Education begins at around six years of age but our learning begins at birth". In fact the capacity of a toddler to learn is many thousands of times greater than an adult's learning ability. Dr Glen Doman also asserts that "Every child born has a greater potential intelligence than Leonardo Da Vinci ever had, with no exceptions!" This should make us all rethink our view of the tiny baby and toddler as not being as simple a creature as we have been led to believe but to recognize the fact we are nurturing a potential genius with a greater capacity to learn than we ever thought possible.

Chapter 5

Is Your Doctor a Health or Sickness Doctor?

A Serious Question for the NHS!

This is a serious question that goes to the heart of our current crisis in the NHS. Most Doctors are trained as pathologists and have very little training in health promotion or natural health therapy. The very word "Doctor" used to mean "Teacher" in the days when physicians had to make do with the few simple remedies at hand and then do their best to understand why their patients were sick and try to motivate them to help themselves as much as possible with good advice. It was in the interest of both doctor and patient to learn how to stay well, as in earlier times even a minor ailment could well prove fatal.

The Foundations of a Sick Society.

We have laid the foundations for an ever increasingly sick society, with the rise of the drug industry driven by profits and increased sales, the ever-increasing focus on pathology in our medical training and practice and the willingness of the public to be passive recipients of whatever treatment is given them, not realizing that they are the prime movers responsible for their own health or sickness. The promises of a golden future for us all, of eliminating all the killer diseases known to man, has turned out to be a lie. I don't believe the blame can be placed at any one group's door but we have all participated in this process either through ignorance, blind belief in this promised future golden age or by design where there appeared to be more money to be made from the sick than the healthy.

The Way Forward.

I believe without a radical change to the way we provide our health care and a massive public education programme to teach personal responsibility for our own health, the current system of medical care will be overwhelmed and unable to provide essential services. This was indeed the essence of a presentation by the Kings Fund at the International Health Workers Conference in Brighton in

2004 that I attended. Unfortunately little has changed since then and the government still feels obliged to pour more resources into our NHS with ever diminishing returns in the way of improved health to society.

Very simply put, more medical care will never lead to a healthier society but to a society increasingly dependent on medicine for survival and a passive population unwilling to take personal responsibility for their own health or ignorant as to how they could do this. Can the general population be blamed when little support or guidance has been given by the medical profession that is ill-prepared for this task, being expert in pathology and disease management but untrained in ethology and health promotion? They too, therefore, cannot be blamed. This is a problem we have all brought on ourselves collectively and individually.

The way forward may be difficult but self-evident; a shift in focus from pathology to ethology, from disease management to health promotion. TotalHealthMatters! is among the thousands of similar natural health centres in the UK, that have contributed to this change for the millions of patients who have sought the help from us that has been wanting in our NHS. When will our leaders recognize the vital need for this in the NHS and provide such a service free or even at a small fee? This would benefit all those who cannot afford private services and also take pressure off the overburdened NHS as well. The beginnings of a Real Health Service.

Chapter 6

What's The Cause of a Disease? Pathology or Ethology?

Many years ago I asked myself the question, "What is the cause of health or what is the aetiology of health?" A much wiser and experienced doctor also asked himself this same question over eighty years ago, he was a pathologist and although he understood the origin of most diseases, he had no answer to this question. He spent the rest of his life researching the question and his work is still as valid today as it was then. Read the Chapter on this entitled "Health is Based on Family and Community"[21]. Seventy years on we find the medical profession is now trying to encourage this approach with "Social Prescribing"

Well, returning to the question of "What is the origin of disease?" We have been taught that most diseases have specific causes, usually just involving one or two, they could be a virus, smoking, a gene or other such factor. This has become more and more confusing for the public as researchers discover links between more and more factors and particular diseases. So perhaps the answer is far more complex than we have been led to believe?

What if, instead of concentrating our research and study in the field of pathology, we chose to shift our attention to more research and study in the field of ethology, or the study of health?

A few years ago I published a small book entitled "Connection – Towards a better understanding of health in medicine"[13] that made the case for the fact that health is connected to practically everything and that health could be regarded as the normal state of affairs. The forward in my book drew on the wisdom of Leonardo da Vinci who combined reductionism and holism in all his work. We need to study the finer parts of any problem (this is the task of reductionism) to help our understanding but we also need to see how the whole system works (this is the task of holism).

Medicine has placed too much emphasis on reductionism to the detriment of understanding the whole, or holism.

So if we are to understand where any particular disease comes from we need to take a far broader inspection of the sufferer's life history to the emergence of their disease.

Just as health depends on many factors including: our physical structural integrity and functioning, our food and fluid consumption, our quality of breathing, our levels of stress, our family and community support, the environmental factors of pollution in our air and water, infections from bacteria or viruses, our genetic make-up, our physical exercise, our mental health, our work, and many more factors. So too, any disease condition will require some or all of those same factors to allow its development.

The good news is that we all have control over the vast majority of those factors, sometimes we may need professional guidance and help and indeed there are such factors as toxins in our environment that are difficult to avoid. Even here, our bodies have the capacity, when health is optimized, to eliminate or nullify the effects of toxic substances.

To summarize I would contend that the greatest protection against every disease known to man is HEALTH!

This is not a crazy oxymoron as it might seem, a health-promoting lifestyle has been shown to offer remarkable resistance against all the major diseases. So perhaps we should all take the optimistic view that by making every effort to improve our health we will give ourselves the greatest protection against every disease and live a long healthy, fulfilling life.

Chapter 7

Health & Fitness, Are They The Same?

There is a lot of interest in keeping fit, getting fit and fitness training systems. But what is fitness? Is it about improving ones health? Is it about developing muscle bulk and strength? Is it about building stamina and physical abilities?

Wikipedia has a definition: Physical fitness is a state of health and well-being and, more specifically, the ability to perform aspects of sports, occupations and daily activities. Physical fitness is generally achieved through proper nutrition, moderate-vigorous physical exercise, and sufficient rest.[26]

My question is "Could a person be physically very fit yet not be healthy?"

We hear many stories of top athletes and body builders suffering strokes, heart attacks or any other serious diseases despite the fact that they had achieved excellent physical strength and stamina. On the other hand I would expect physical health for any person, would be an essential part of being healthy in mind and body. The confusion lies in the latter part of Wikipedia's definition in my opinion; "nutrition, exercise and rest".

How many fitness centres and gyms promote the increased intake of quality meat for protein, the increased consumption of nutrient supplements such as milk protein, vitamins and minerals without offering adequate advice on what might constitute optimum nutrition?

There are many myths that are dispelled by solid scientific research, they include the belief that to be fit and strong we need quality protein from meat sources, that we couldn't thrive on just a plant based diet, that vitamin and mineral supplements are good for providing all the body's needs and that with physical fitness comes improved health and increased longevity.

Taking each myth in turn, it is a simple fact that the protein content of meat is about 30 grams per hundred calories consumed, and for plant-based food it is the same 30 grams per hundred calories. To rely primarily on an animal-product diet for one's protein needs would lead to a severe deficit of micro-nutrients such as vitamins and minerals, for the simple reason that the animal has already used up these nutrients for itself and with the consumption of the animal products come added toxins, residual antibiotics, or growth hormones generally found in factory farmed meat.

As regards the use of mineral and vitamin supplements to make good such deficits, we discover things are a bit more complex, the body cannot process and absorb efficiently individual vitamins or minerals without the many other complex phytonutrients found in whole plant food. For instance to replace the vitamin C benefit of an apple for our nutrition, we would have to consume many more times the quantity of vitamin C as a supplement, most of which would not be absorbed.

Finally although no one could disagree that we need a physically healthy body to enjoy optimal health, by over-concentration on this to the detriment of our diet, sleep, rest, lifestyle and mental state we can lay the foundation for other serious health problems including heart disease, diabetes, hypertension and cancer.

So my recommendation is take a more holistic approach to your health and wellbeing, yes, exercise, especially if your work is desk-bound, but don't neglect your diet, your stress management, your breathing, your sleep quality, and get advice on all other lifestyle factors that will improve your health and longevity. As my book "Connection" suggests and develops, health is connected to ....everything!

Chapter 8

What Is Health? Is It More Than an Absence of Disease?

Whenever I have asked an audience this question most people intuitively know that health is far more than freedom from diseases. Although health is a very difficult concept to define we do know what are the main causes of good health.

If you have ever wondered what are these key factors that determine our health and longevity and would like some practical, reliable, information on this subject then read on.

It is unfortunate that modern medicine has been primarily concerned with disease, illness and pathology for the past century and has not given the public much advice on health promotion. The main reason for this omission, is that doctors are not trained as health practitioners; they receive very little training in the fundamentals of health promotion: the need for good body mechanics, for good nutrition, for adequate exercise, good breathing, good relaxation, a healthy environment and a supportive community.

Because of this disease model of medical practice, the system is overloaded with a sick population that is creating increasingly impossible demands on finance and resources.

We also have trained an entire generation of the public to rely on their doctor to keep them well and have lost the sense of self-responsibility for their own health or are just unwilling to do anything to help themselves. "Responsibility" is exactly the right word as it literally means, "the ability to respond to any change", we all have this innate ability but may have been discouraged from using it when it comes to our own health care. This personal response-ability has been usurped by the medical profession and by all therapists that fail to help and teach their patients to help themselves, but make their patients increasingly reliant on their therapy support. There is today a prevalent view that "It's pointless worrying about our own health and quality of life or longevity, just carry on and enjoy yourself as we all have to die of something don't we?" The truth is we can have a profound effect on our health and longevity and we can usually choose either a long healthy, active life or a chronically sick existence for many years maintained with increasing medical & surgical interventions.

If we follow the American lifestyle over the next few years we can expect the same outcomes; according to good medical authority we may be seeing parents routinely outliving their children for the first time in our history. It is now commonplace in America for young people to be suffering strokes or heart attacks and other chronic degenerative diseases that were usually found among the middle or old aged in the past. The major reason for this catastrophic decline in health and increased premature death appears to be largely due to diet, lack of exercise & an unhealthy lifestyle with increasing medical interventions with drugs and surgery.

After thirty-five years in the healthcare profession I have tried to find an answer to this question from past doctors who have devoted time to research this topic. A few years ago I decided to put all my findings into one small book called "Connection – Towards a better understanding of health in medicine."[13] The conclusion I came to was that our health depends on almost everything! This may have been the reason doctors have preferred to follow the path they have taken, trying to fix or ameliorate diseases rather than trying to teach their patients to keep themselves healthy.

However there are just four factors we can all control ourselves, that can go a long way to promoting health and reducing the risk of developing serious chronic diseases. Let's look at each of them now:

Our Body.

We must not forget that our body is an amazing mechanical structure that can suffer many of the problems that might affect any machine. It needs to be kept in fine adjustment to function optimally.

We wouldn't dream of driving a car with a twisted chassis or faulty wheel alignment, but most people happily try to get around with many mechanical problems in their spine. Luckily this is where your body is radically different from your car; it is able to compensate for many mechanical faults but at a cost of either pain, poor functioning or long term damage needing replacement hips, knees or organs. The first chapter of "Connection" explains the details of our body mechanics, the relationship between problems in our spinal structure and impaired functioning of related organs or tissue, and what we can do to help ourselves.

Our Food.

We are what we eat, or "let food be thy medicine and medicine be thy food". Our diet is the most important factor underpinning health or laying the foundation for serious diseases for most of us in the West. I would heartily recommend you read an amazing best - selling book giving the research behind this concept it is entitled "How Not To Die" by Dr Michael Greger.[29] My book "Connection" gives a very brief summary of some of the research findings that have led to these conclusions. Unfortunately the diet and nutrition industry has totally confused people and is driven by commercial interests that are very much like the drug industry.

The truth seems to be that the most health-promoting diet *is the simplest and cheapest* to follow. It is described as Whole Plant Nutrition[30] and recommends you eat all you want of whole plant based foods and dishes. You will not need supplements (with just two possible exceptions), enzymes, special training or calorie counting. This diet drastically reduces the risk of developing most of the chronic diseases, normalizes your weight and will often reverse pathologies such as type 2 diabetes, heart disease and hypertension. This same diet, as a bonus, may also be the fastest way of tackling global warming and the destruction of our planet if most of the world's population followed it.[31]

On the subject od nutrition, remember most of us don't drink enough water for our health.

Our Breath.

Your doctor checks your blood pressure, your weight, your reflexes, your blood chemistry, your heart & lung sounds, you may have had Peak Flow tests, even given you X-rays or MRI scans but why didn't he check your breathing rate?

We can live for three weeks without food, for three days without water but only three minutes without air. That life's most fundamental activity, BREATHING, is not routinely checked by Western doctors (unless you are complaining of breathing difficulties like asthma, COPD or bronchitis) is a serious omission.

If there were just one measure allowed when checking the health of a person I would suggest THEIR BREATHING would be the best choice. Why?

Firstly it's a quick, easy and cheap test. It takes less than a minute, is none invasive and does not require special expensive equipment.

Secondly, many medical researchers have claimed dysfunctional breathing may be the cause of over a hundred modern diseases, so wouldn't it make sense to at least diagnose it and then perhaps address this problem first?

Thirdly, although our breathing is normally automatic and unconscious – like our heart, liver, & kidney activity - we can take conscious control of it and change our bad breathing habits for good normal ones, with just a little training.

Fourthly, this route to better health and wellbeing empowers the patient and reduces their reliance on medical care and drugs. Surely a very desirable aim?

My third chapter of "Connection" explains how to check your own breathing and how most diseases are associated with poor breathing. You may decide to study my free podcast breath-training course. "Better Breathing Means Better Health"[32], found on all major podcast sites.

Our Mind.

Did you know that your brain tomorrow will be different from how it is today? This is the latest thinking of how our brain functions; it is about "the plastic brain". The most amazing thing about the human brain is that it has the ability to grow, change, re-arrange all our thoughts and mental processing from birth to death. We have a plastic brain that can do remarkable things such as cope with a stroke and re-organized all the activities that were done in one area that is damaged and allow them to be done in another area. Areas of the brain that are normally associated with speech can be used for sight or hearing. The old thinking was that we were born with a brain divided up into little boxes, each was allocated a specific job, this has all changed with the latest research.

What does this mean to you and me?

The first lesson we need to learn is "never to say never!" We cannot tell what is possible and what is not possible. The second lesson is that we can change the way we think, feel or behave in remarkable ways with a little help and effort. The third lesson is that despite the fact that we function much of our waking time with habits learned since childhood (it would be tough to have to learn how to talk, walk or clean our teeth the new every day!) all these habitually programmed activities can be re-programmed. How we can do this is the subject of this section of my book. The phrase used today is "If it fires, it wires!" Repeated nerve signals create stronger pathways. This is well explained by Norman Doidge in his book "The Brain's Way of Healing"[33] Did you know that every child, at birth, has a greater potential intelligence than Leonardo da Vinci ever had, that includes your children and grandchildren!" This finding is based on many years of clinical research and practical application by the mothers or fathers of tens of thousands of babies guided by Dr.

Glen Doman and others. This should make us all rethink our view of the tiny baby and toddler as not being as simple a creature as we have been led to believe but to recognize the fact we are nurturing a potential genius with a greater capacity to learn than we ever thought possible. Watch the video on YouTube entitled "Glen Doman Introducing The Institutes (Part 1), Working With Children for over 50 Years"[8]

Finally, remember your mind is closely linked to all of the body activities and is perhaps the most significant connection discussed in my book. We are emerging from a long period of study and thinking based on reductionism, or the analysis and study of increasingly smaller and smaller parts of our whole, but now there is a move towards the more difficult task of understanding how every part is connected to every other part and every system is affecting every other system, this is Holism or Wholism. We must remember we need both reductionism and holism if we are to make any sense of the universe we live in.

My book "Connection" discusses other key factors related to our health including family, community and even cosmic factors! It also offers an alternative theory of the universe to deal with the awkward question "What happened before the Big Bang?"

Chapter 9

Chronic Hidden Hyperventilation – 21st Century Epidemic

It is estimated that 90% of the population in the West are in the habit of over-breathing, they suffer from chronic hidden hyperventilation. Because chronic hidden hyperventilation is not easily recognized, it is rarely diagnosed, and when it is, patients are given little advice and support to deal with a habit. The Buteyko Method of breath training appears to be the most effective system to correct the problem.

Why is it termed hidden?

Normal breathing at rest involves breathing about five to six litres of air per minute with a respiratory rate of between eight and twelve breaths per minute. If a person breathes slightly faster say between twelve and sixteen breaths per minute, this will not be easy to noticed by any observer, but that will increase their minute volume by two to three litres. If they breathe slightly larger breaths, perhaps 25% larger, this will increase their minute volume by other two to three litres, but again this increase will not be easily observed. If they habitually sigh or yawn this can add another two to three litres per minute. The total effect will be to increase the volume of of air breathed per minute to ten to fifteen litres, two to three times our normal requirement.

Chronic hidden hyperventilation is frequently associated with habitual mouth breathing as it is far easier to breathe heavily through the mouth than the narrow passages of the nose. The effects of chronic hidden hyperventilation is to lower carbon dioxide levels in the body, to increase the risk of airborne infection, as the filtering protection of the immune system in the nose is by-passed, to irritate and inflame the airways as dry or cold air is breathed through the mouth and impair the immune system and risk constriction of blood vessels because of loss of nitric acid, produced naturally in the nasal cavities. The lower carbon dioxide disturbs the entire physiology of the body; the blood does not release its oxygen to the tissue freely due to the Bohr Effect. Smooth muscle wrapped around all hollow organs of the body, airways, arterial blood vessels, bladder and gut is caused to spasm restricting circulation and breathing and causing digestive problems. The body's pH shifts to becoming more alkaline and this affects every biochemical activity in the body.

So do you hyperventilate?

Check your own breathing now, if you have a CP or control pause (the Buteyko Measure of your breathing) of less than twenty seconds this is the simple way to improve your health and fitness in just a few weeks. The health consequences of chronic hyperventilation can be serious. Conditions invariably associated with this problem include asthma, sinusitis, breathlessness, angina, diabetes, hay-fever, low-energy, gut problems, hypertension, chronic fatigue syndrome, panic attacks, snoring, IBS and much more. Improved breathing invariably helps all these conditions and improves wellbeing.

Breathing exercises aimed at normalizing breathing volume provide therapeutic benefits, it takes less than a minute to measure your breathing quality. The measure I recommend is called Control Pause, the key monitoring tool of the Buteyko Method developed by Professor Konstantin Buteyko. It is a measure in seconds of your maximum comfortable breath hold after exhaling while at rest. It is a useful guide to the degree of hyperventilation and the level of carbon dioxide levels in the lungs.

Check Your Breathing Now

Take a breath in through your nose, keeping your mouth closed all the time, breathe out through your nose, gently hold your nose and your breath until you feel the need to take a breath in, then release your nose then take a normal breath in through your nose. With a timer note how many seconds you held your breath, this is your Control Pause.

If your control pause is less than 20 seconds you would be well advised to improve your breathing as you are severely over-breathing and this will take its toll on your health and fitness, if you are breathing normally you should have a control pause of between 45 and 60 seconds. Most patients I check have a control pause less than 30 seconds, many who have health problems will have a control in the teens. With the benefit of the internet you can learn to improve your breathing in just two or three weeks anywhere in the world with our Skype Course[6].

Chapter 10

Half a Trillion Breaths

It seems every mammal is allotted the same number of breaths in their lifetime, roughly 600,000,000. Some whales and certain elephants manage on as few as four breaths per minute and can live to 150 years whereas the busy fast moving pigmy shrew who breathes around 500 breaths per minute lives just over a year.

The table below shows the data for other mammals including us, man.

Mammal	Breaths per minute	Lifespan (Years)
Whales	3-5	200 - 130
Elephant*	5-12	130 - 65
Man	6-16	110 - 45
Horse	12-15	55 - 45
Sheep	16-34	40 - 20
Cat	20-40	30 - 15
Dog	20-30	20 - 15
Cow	25-50	25 - 15
Pig	30-60	20 - 10
Hamster	35-135	10 - 5
Mouse	90-250	7 - 2
Shrew	300 -800	2 - 1

* Elephants breathe 4-6 breaths/min lying down & 8-12 standing, they have a different lung structure.

"The perfect man breathes as if he is not breathing" Lao Tzu (4th century BC)

Lao Tzu is claimed to have lived to a 160 years old, but this is generally regarded as a myth, I tend to disagree for if he only breathed about five breaths per minute, like an elephant or a whale he could well have lived that long.

"The more you breathe the closer you are to death. The less you breathe the longer you will live." Konstantin Buteyko1923-2003.

We don't promise great longevity when you train with the Buteyko Method but you will have better health, more energy, sounder sleep, fewer symptoms and a calmer life if you breathe better.

With humans, one of the major factors that cause chronic hidden hyperventilation is stress. Stress triggers the primitive fight/flight response repeatedly, eventually causing the carbon dioxide receptors to accept and re-adjust to maintain a lower level of carbon dioxide and thereby establishing a over-breathing pattern.

Your doctor usually never checks your breathing as part of a routine examination (unless you arrive complaining of a respiratory condition) despite the fact that breathing is perhaps the most important activity in our lives! Well, we can live for three weeks without food, three days without water but less than three minutes without air! Over 75% of us in the West over-breath or hyperventilate and breathe badly using our upper chest and mouth breathing instead of using our amazing "breathing tube" – our nose. I try to get this message across to young children by telling them, "You should breathe through your mouth as often as you eat through your nose!" Your nose is for breathing, your mouth for eating and talking!

Chapter 11

Mindfulness and Your Breathing.

One of my inspiring teachers, a Dr Scott-Williamson of the Peckham Experiment fame, said we all live in three states throughout our lives; the first is living, the second is surviving and the third is dying.

By living, he meant living healthily, intensely and fully, enjoying or experiencing every waking moment, the kind of living most toddlers enjoy, where every day is a wealth of new discoveries, experiences and magic.

By surviving, he meant following whatever routine allows us to get through the day, working, eating, sleeping and doing whatever is necessary to make this possible.

By dying, he meant the slow aging through the passage of time towards our end of life, a kind of suspended animation.

He suggested we should all assess what was the mix of these three states in our own lives. A rich, happy and fulfilling life might be made up of a mix of 70:20:10, but an unhappy, aimless life would be nearer a mix of 10:20:70.

Another way of assessing this would be to check how much time we spend living in the past or worrying about the future as opposed to living in the present. *The reality of time is that, the past no longer exists and neither does the future, all that is real is this immediate moment, the NOW! In fact there are an infinite number of "now" moments throughout every waking day and only when we are very young do we live in them most of the time.* A special treat promised tomorrow or even later in the day has less impact on the child than what is happening right there and then. What happened yesterday has largely been forgotten as today's discoveries take all their attention.

However, as we grow older and not always wiser, much of our lives are concerned about what may happen tomorrow or next year and memories of all that happened yesterday or last year cloud our perception of the present.

Sages have for thousands of years taught that we should all learn to be more mindful and aware of the present and there has been a growing new interest in "mindfulness" as a way to a happier life.

The problem with trying to be mindful all our waking hours is that it requires the mental discipline on a Buddhist monk! What gets in the way is the stress of our modern lifestyle that turns us into chickens running round without their heads. If we could be calmer throughout the day despite the stressors we would naturally become more aware and mindful of all around us.

It is interesting to note that all the teachings of mindfulness include breathing exercises and practices, and since our breathing is controlled automatically without any conscious effort on our part, perhaps another way would be, to re-learn how to breathe normally by correcting our automatic system.

The stressful person is always breathing faster and deeper than necessary, this is because of our primitive "fight or flight" response to any emergency or danger, as if in a state of high alert, as if being attacked by a predator. This is not a state to be in if we want to be fully aware and mindful, it is a critical survival state. If we could change our automatic breathing to be always quiet, calm and gentle then we would be also be in a quiet, calm and relaxed state ourselves.

Our automatic breathing is controlled by receptors in the brain that measure our carbon dioxide levels and adjust our breathing rate to maintain a constant level. Because of habitually being stressed and over-breathing the receptors get used to the low level of carbon dioxide and from then on keep us over-breathing at this stressful state.

It is almost impossible to be stressed if our breathing is calm.

The good news is that this is exactly what we can correct with the Buteyko Method of breath training. Students are taught how to effectively reset the receptors in our brain to return our breathing to normal. Once reset they stay that way and we are then better able to meet all the daily stressors with a calmness that allows us to be more mindful naturally, without the need to do specific exercises.

Finally, remember your mind is closely linked to all of the body activities and is perhaps the most significant connection discussed in my book, "Connection". We are emerging from a long period of study and thinking based on reductionism, or the analysis and study of increasingly smaller and smaller parts of our whole, but now there is a move towards the more difficult task of understanding how every part is connected to every other part and every system is affecting every other system, this is Holism or Wholism.

We must remember we need both reductionism and holism if we are to make any sense of the universe we live in. My book "Connection" discusses other key factors related to our health including family, community and even cosmic factors! It also offers an alternative theory of the universe to deal with the awkward question "What happened before the Big Bang!"

Chapter 12

Your Child's Crooked Teeth, Their Breathing & Their General Health

Major medical breakthroughs can come from the most unexpected places, who would think an artist could transform orthodontic practice?

George Catlin, an American artist born in 1796, is famous for his remarkable record in paintings and notes on native American Indians, but his great genius has been almost ignored till recent years. In his book entitled "Shut Your Mouth & Save Your Life"[3] written in 1870 he details his assertions that the bad habit of modern man of mouth breathing was the cause of much of his disease and disfigurement. This was based on his close observation and questioning of thousands of native Indians & white immigrants. It has taken almost one and a half centuries for modern research in medicine to recognize the validity of this concept. Today there is a growing number of specialist orthodontists and health workers who say the same thing based on sound scientific evidence.

On 5th October 2012 there was a Conference in New York of American Association of Physiological Medicine & Dentistry discussing this association. Since then an increasing number of orthodontists are adding breath training to their regular practice services especially with young children whilst their cranial structures are developing. So now we know that orthodontic problems like crooked or crowded teeth are not due to genetic factors but to the bad habit of mouth breathing. You can watch an excellent video on YouTube by the leading international dentist Dr John Flutter who explains the technicalities, just search the internet for Dr John Flutter.[4]

If the only adverse effect of mouth breathing was poor facial development that would be bad enough, but mouth breathing also predisposes children to other health problems including: recurrent throat infections, enlarged tonsils, asthma, poor sleep and much more simply because their bodies are poorly oxygenated and the protection against infective agents is lost by not breathing through the nose. Everyone can learn to breathe normally in just a few weeks with the Buteyko Method Training. Much of the success of the Buteyko Method of training is due to the elimination of mouth breathing, advice against over-eating, the encouragement of more physical exercise and advice on quality sleeping.

The ideas are so simple to teach or learn that they have been dismissed by mainstream medicine that increasingly puts its faith (yes, I use the word "faith" intentionally) in drug therapy or other intensive medical interventions.

The truth is there is little profit to be made from simple remedial systems that could threaten the profitability of our international pharmaceutical companies and the vast industry built on the management of disease, if the public were better informed of their existence. Meanwhile take Mr. Catlin's advice and ensure you and your family "Shut Your Mouth & start on the road to better health!"

Chapter 13

What is Asthma? – Big Business or Medical Confusion?

This question was posed in the Lancet over twelve years ago (Vol. 368, No. 9537, 26/8/06) and demonstrated there is no consensus to the answer. To add to the confusion recent research suggested that up to 40% of those diagnosed with asthma had been wrongly diagnosed and should not be on the medication prescribed. (Middlemore & Green Hospitals March 25th 2007)

This must seem confusing for all those suffering the many symptoms that constitute the diagnosis of asthma. The wheeze, the breathlessness, the high mucus production, the acute allergic reaction to many triggers, the irritating cough, the low energy, the disturbed sleep that all appear to be relieved with prescribed asthma medication. Surely with this array of symptoms we can rely on an accurate diagnosis of the condition?

There is no doubt that modern medication helps manage the symptoms associated with asthma, but what if the underlying cause of asthma symptoms were something as simple as dysfunctional breathing?

This was exactly the conclusion that Dr Konstantin Buteyko came to after forty years of careful research. He discovered that every asthmatic over-breathed and that their asthma attacks occurred when their over-breathing was excessive. He discovered that if patients were taught to normalize their breathing and make minor adjustments to their lifestyle, their symptoms would disappear and would no longer need medication. His teaching came to the West initially in Australia in the 1980's and has spread across the world since then, much to the relief of hundreds of thousands of asthma sufferers who now either need no medication or far less than they used to need. But here lies the major problem.

The medical profession rightly demanded more research papers to verify the Professor's findings and the experiences of all those happy asthma sufferers who had supposedly benefitted from their breath training.

In fact all the clinical trials of the Buteyko Method for the support of asthma patients have demonstrated up to 90% reduction in the need for reliever medication and up to 50% reduction of steroids, reduced coughing, reduced wheezing and less breathlessness along with improved sleep and general quality of life. However additional large trials would need financing and the usual source of funding of clinical trials is from the drug companies who benefit from the sales of asthma drugs, often around twenty per cent of their earnings come from this product group. They reasonably have not offered to fund such trials, as the confirmation of this relationship would severely hit their profits and their first responsibility, as for any company, is to their shareholders.

You may want to see two videos on my website entitled "Understanding Asthma" and "Understanding Asthma - A Different Viewpoint" that compare these two views of the origins of asthma.[24]

Meanwhile millions of people are being treated with ever increasing doses of asthma drugs with poor results. The UK is one of the worst countries in Europe for asthma treatment according to the latest reports. Why should this be? There are many reasons besides the fundamental one of treating symptoms rather than the root cause. Inappropriate prescription of asthma drugs, inadequate A&E support, poor compliance of patients with their asthma management programs and excessive reliance on medication rather than greater education and training of asthma sufferers have all led to the current situation.

The reason for our failures is being put down to inadequate funding, but more cash is not always the panacea.

If every asthma nurse were to teach their patients the significance of breath control based on the clinically proven Buteyko Method there would be two major outcomes; the asthma drugs bill for the NHS would be reduced by half and patients would be better able to manage their asthma with less reliance on drugs and enjoy a better quality of life.

The cost of the additional training would be offset within months by the ensuing savings on drugs, doctors' appointments and reduced A&E admissions. Two doctors who referred a number of their asthma patients for Buteyko Training found they were saving thousands of pounds on drug prescription and medical intervention; the cost of training was quickly offset by the savings in the first year and since the training is a one off, the savings continued into the future.

At present Buteyko Training is only available from private centres but considering a course may cost less than the servicing of your car and could transform your life, this ought not to be a barrier to asthmatics paying themselves. To make this approach more available to the growing population of people diagnosed with asthma I have produced the full training programme as a podcast.[25] This is not ideal as one-to-one training is most effective, but it will give significant help to the many who for whatever reason cannot enrol on a face-to-face Buteyko Training Course.

Chapter 14

Diet v Breathing Relationship

This article proposes that what we eat is strongly connected to the way we breathe, and that the way we breathe has a profound impact on the way we eat. Chronic hidden hyperventilation CHHV or over-breathing is related to stress, diet and bad breathing habits, but diet appears to be the major factor, perhaps because a stressful lifestyle usually leads to bad eating habits, as well as directly affecting breathing due to the fight/flight responses to stressors. Over the past three years I have gathered data on breathing and diet from every patient and the interim results are shown below:

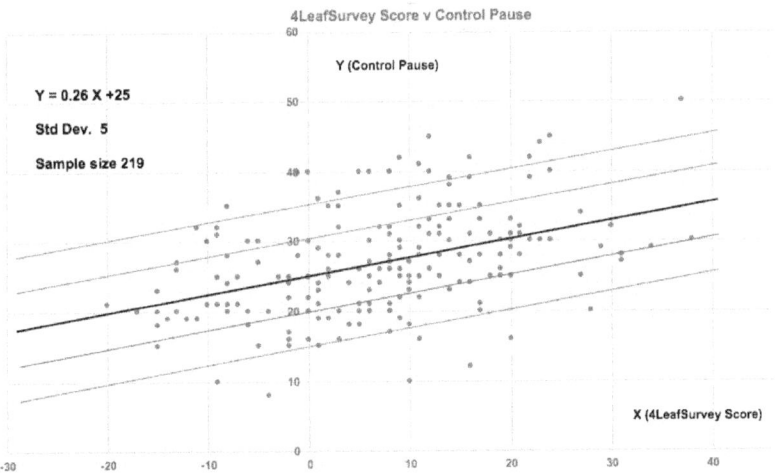

The above graph is based on 219 data points from patients, their Diet Score is an estimate of the % age of calories derived from Whole Plant Food and their Breath Score is an estimate of their body oxygenation or degree of chronic hidden hyperventilation based on the Buteyko Method Control Pause.

The above graph is based on 191 data points. Y = 0.26X + 25. Standard deviation = 5

Diet measure used: The 4LeafSurvey that estimates the %age of calories derived from Whole Plant Foods (Ranges from -40 to +40) (www.4leafsurvey.com)

Breathing measure used: The Control Pause that estimates the degree of Chronic Hidden Hyperventilation (CHHV) or the %age of CO_2 in the lungs. (Ranges from 0 to 60 +)

4LeafSurvey Score (X)	Control Pause Score (Y)	CP Range
- 40	15	10 to 20
- 35	16	11 to 21
- 30	17	12 to 22
- 25	19	14 to 24
- 20	20	15 to 25
- 15	21	16 to 26
- 10	22	17 to 27
- 5	24	19 to 29
0	25	20 to 30
+ 5	26	21 to 31
+ 10	28	23 to 33
+ 15	29	24 to 34
+ 20	30	25 to 35
+ 25	32	27 to 37
+ 30	33	28 to 38
+ 35	34	29 to 39
+ 40	35	30 to 40

It takes less than five minutes to have clients complete the 4leafsurvey and establish an estimate of the % age of calories they derive from whole plant food.

As I have explained in previous articles, it is important to have a measure of a client's diet before they are advised to make major changes in their breathing, especially if they are eating a very acid-forming diet that will present as a 4leafSurvey Score with a high negative value.

This was not as important when Professor Buteyko did his research in Russia, as the typical diet was radically different from the Standard American Diet (SAD diet). Nevertheless he did establish the fact that certain foods could increase CHHV and advised patients to abstain from eating them when ill. Today, in the West that list would be much longer, as our diets have become increasingly based on a heavy consumption of animal based foods, dairy and convenience processed foods.

There are many instances where patients in the USA have changed to a whole plant diet and their asthma has totally cleared. Hypertension and many gut problems have responded quickly to such a dietary shift. As CHHV is a 21st century epidemic in the West, there is no doubt that breath training is usually needed to break old breathing habits that may have arisen from stress, trauma, illnesses, mis-guided efforts to do deep breathing as well as lack of physical exercise and dietary excesses.

The potential danger of having clients eliminate CHHV who are eating an acid-forming diet could be depletion of bone calcium or development of kidney stones as the alternative buffering by increased breathing and elimination of carbon dioxide is taken away. I am sure that in time the improved breathing would result in more mindful eating habits, but why not help the patient with advice on their diet as well as teaching them to breathe better? The evidence is now overwhelming that the SAD diet is a direct precursor to heart disease, hypertension, diabetes, obesity, cancer and many other modern chronic diseases. If we are serious about helping our clients to better health we should at least pass this information on to them; whether they decide to make any changes in their eating habits, that is their decision, but a decision they will make with better information that they will not get from most doctors.

My training was as an osteopath, and although osteopathy was originally meant to be a total system of healing, baring surgical interventions, I do not think there are many osteopaths who have the same dedication and understanding as our earlier founders to offer patients sufficient care with just physical therapy. It was for this reason that I trained as a Buteyko Educator and Plantrician and now give advice and support on diet and general lifestyle matters.

I would strongly encourage all Buteyko Educators to make themselves acquainted with the China Study and the Whole Plant Nutrition diet and to introduce the on-line 4leafsurvey to their clients as a routine. Even better, why not take the excellent on-line training offered by eCornell University to gain a Whole Plant Nutrition Certificate? For me it has been the best investment in health education I have ever made.

Chapter 15

Hypoxia & Its Treatment

Hypoxia, or inadequate tissue oxygenation, can occur in many ways but the effect on the healthy functioning of the affected tissue is usually the same, poor functioning in every way. Perhaps the key problem arising from tissue hypoxia is impaired energy production by the cells' mitochondria. This reduced cellular energy lays the foundation for an unhealthy cell and disease.

The evidence is now well established to link most chronic disease to hypoxia, this includes heart disease, diabetes, cancer, and almost every other disease of modern man.

So how can we combat hypoxia?

There are two proven approaches to improving tissue oxygenation. One involves improving the way we breathe and eat, the other uses oxygen-enriched water.

Breath Training to Alleviate Hypoxia

Poor breathing or more specifically chronic hidden hyperventilation is a 21st century epidemic in modern societies, arising from the lifestyle of modern western man; stress and diet being the two main factors causing this, stress and diet.

Stress triggers the primitive fight/flight response that includes increased respiratory rate. Chronic hyperventilation leads to loss of carbon dioxide and subsequent poor delivery of oxygen to the body's tissue.

Excessive over-consumption of meat, dairy and over-processed junk food leads to metabolic acidosis that the body tries to correct by eliminating large amounts of carbon dioxide by hyperventilating.

Usually both these factors are involved. My recent research has shown a high correlation between the way we breath and what we eat, read Chapter 14.

Thus long-term health benefits can be achieved by learning better breathing and improved eating habits.

The Buteyko Method of breath training improves tissue oxygenation long-term, combined with a shift to a Whole Plant Diet that eliminates metabolic acidosis associated with the SAD (Standard American Diet) not only improves oxygenation but gives many beneficial effects from the reduction of meat, dairy and highly processed foods.

As part of any lifestyle changes efforts should be made to ensure adequate hydration. The vast majority of the public today suffer from some degree of dehydration. Remember we are water body creatures with liquid conductors of electric nerve impulses, for more on this topic read Chapter 27.

If there is poor circulation of blood in any area of the body cells *there may still be suffering from hypoxia even with improved respiration and normal oxygen delivery to the rest of the body.*

This is especially the reason for using the second approach of oxygen-enriched water, since every cell must have water for its survival and if extra oxygen can be carried in the water we drink, this will ensure reduced hypoxia throughout the body.

Hydration with Oxygen Enriched Water

To speedily address hypoxia it is now possible to achieve this by drinking oxygen enriched water and even bathing in this water. Because our body circulates water to every cell, if oxygen enriched water is used, every cell in our body is better and rapidly oxygenated, hypoxia is eliminated, even in areas where blood circulation is impaired, where improved breathing has less impact.

The oxygen enriched water is not simply water with added oxygen but it is the result of a special treatment that extracts hydrogen from the water leaving highly active and readily available oxygen. It is available internationally under two brand names, Kaqun Water [46] or Elo Water.

The great advantage of this latter approach is that it does not demand any lifestyle changes, re-training of our breathing habits or changes to our eating habits. Improved oxygenation is achieved almost regardless of the individual's current lifestyle.

What appears to happen is that over time, as the individual's health improves, they tend to change their lifestyle towards a healthier way. This is similar to the effect of better breathing leading to better eating that my research has demonstrated.

Practical Considerations

For anyone not suffering from any chronic disease and just wanting to improve their general health and wellbeing it would make sense to combine these two approaches initially and then to reduce the use of the oxygen enriched water over time. Such an approach would give a kick start towards improved oxygenation and all the health benefits that derive from this. In the real world, the additional cost of oxygen-enriched water must be taken into account.

For others with significant ill health associated with hypoxia, a more intensive use of oxygenated water for longer periods would be better combined with lifestyle changes as and when they felt ready. If the hypoxia has been causing damage for some time, a longer period of such water support would speed up repair and normalization of tissue oxygenation. Improved oxygenation may reduce the need for other medical interventions that would offset the additional cost of this approach. Further research is in hand to demonstrate the remarkable benefits of this approach, clinics are already providing such therapy with regular bathing in this water alongside drinking it. Early results have been very encouraging especially for the beneficial impact of reduced hypoxia with cancer and diabetic patients.

Chapter 16

Beautiful & Healthy Children & Adults
Eat Starch & Nose Breathe

It is always fascinating when two pieces of a puzzle fit together; what has nose-breathing and eating starch got to do with our health and beauty?

Slowly I am coming to the conclusion that everything connects with everything! Over 150 years ago an American artist, George Catlin observing the native and immigrant populations in North and South America, noticed how well-formed and fine-featured the indigenous people were and how ugly and sick many of the immigrants seemed. He decided it was because the American Indians never let their children breathe through their mouths, always teaching them to nose breathe, whereas it was usual to see the immigrants mouth breathing. The immigrants often had crowded or misaligned teeth and poor facial features. He wrote a book on the subject "Shut your mouth and live a healthy life"[3]

It has taken 150 years for modern medicine to recognize he was right. Now there is a growing group of orthodontists who rarely extract teeth or use braces with young children but teach them facial muscle developing exercises and to breathe correctly, through the nose. They get good results! The bonus is that these same children will be healthier because of their improved breathing, are less likely to suffer asthma, panic attacks, hay fever, skin problems etc. because they are not mouth breathing.

The rationale underpinning these results is easy to understand. When we mouth breathe our tongue lies on the bottom of the mouth, when we nose breathe it presses up on the upper palate 24/7 maintaining a wide dental arch as well as encouraging normal lower jaw development. Dr John Flutter has produced a valuable video explaining this relationship.[4]

Let's start with a simple fact you can check with your dentist or orthodontist; a pressure of 3 grams applied 24/7 to a tooth will cause it to move to a new location – this is how braces work. When we nose breathe the tongue presses on the upper palate and applies some 30 grams pressure all of the time while you are nose breathing.

Test this out for yourself!

Now begin the experiment: open your mouth and breathe through your mouth for a minute. Note mentally where your tongue is.
Close your mouth and breathe through your nose for a minute and once again mentally note where your tongue is now.
Unless you are a rare exception you will have discovered that your tongue rested on the lower teeth when mouth breathing but pressed up onto the roof of the mouth when nose breathing. When pressing on the roof of the mouth it is applying about 30 grams of pressure that keeps the upper jaw wide to comfortably hold all the teeth in good position, if the tongue falls to the lower jaw the upper jaw will usually narrow and the teeth will crowd and be crooked. This, in turn, affects the development of the whole facial features.

Now about the starch!

The human body is designed to eat starchy foods, all societies that get most of their nutrition from starch-based vegetables are almost free from obesity and many of the diseases of modern affluent societies.

Whether it is rice, potatoes, wheat, corn or whatever, they all are the great staples of the large proportion of the world's population who enjoy good health and fine bodies.

The increased proportion of animal based food and refined foods are the downfall of modern societies, giving rise to obesity, cancer, heart disease, strokes, arthritis, Alzheimer's and more. For the science & detailed explanation read "The Starch Solution" by Dr. John A. McDougall[5]

So for you and your children: always nose-breathe and enjoy the carbs (unrefined, don't forget), fruit, vegetables and grains, and eat less of the food derived from highly processed foods and anything with a face.
Reduce your consumption of C.R.A.P foods (Calorie-Rich And Processed foods) that generally have high fat and sugar content, low fibre, minimal micronutrients and potentially unhealthy additives)

Chapter 17

What is A Plantrician?

Although I have been in practice as a therapist for over thirty-five years I only began including nutrition screening and advice three years ago. This was not because I didn't believe nutrition to be one of the most important factors underpinning good health, far from it, I had decided twenty years ago that health was founded on good structure, good diet, good breathing, good mind as well as a good supportive environment from birth, but I had not been able to find any substantial scientifically based nutritional advice to give my patients.

Only when I read "The China Study" by Dr. T Colin Campbell[9] that detailed the research that began with the largest epidemiological study in nutrition ever conducted. This was the information I had been looking for all those years, superbly presented and backed by many years of clinical research as well.

My next step was to enrol on an intensive eCornell University Course over the internet called " Whole Plant Based Nutrition"[10]. Once again I was given an educational feast from the leading doctors and practitioners in this field.

Very soon afterwards I added "The Food Connection" to my website and began introducing this work to all my patients. By this time I had been in close communication with the founders of the 4LeafSurvey[11], Dr Kerry Graff & J. Morris Hicks, exchanging my research on the connection between diet and breathing, and as one of their first practitioners to be licensed, began using this simple but effective questionnaire routinely with every patient.

Since then I have been doing whatever I can to promote the use of this simple screening for doctors and other health workers who want to offer sound nutritional advice to their patients as part of their practice service.

It takes just a few minutes for the patient to complete the questionnaire, just one minute for the practitioner to mark their results with a guide to changes to be made over time, backed up with an A4 information sheet plus, in my practice, access to a substantial library of cook books the patient may borrow, giving them the guidance they need to get started towards a better nutritional lifestyle.

What I find so rewarding is that I now know I can help every patient I see with sound nutritional advice that is simple for everyone to understand but not for some so easy to follow. Whether they decide to make changes in their eating habits is their decision, but at least they now have the information how to change for the better if they so choose.

Only two nutritional supplements may be suggested, B12 if they make all the advised changes to their diet or if blood tests indicate a need for this supplement, and Vitamin D in winter for some, again if there is any concern, a blood test will clarify whether this is needed or not.

Any patient on medication from their doctor is advised to have this checked out as they make substantial changes to their diet. Usually improved nutrition leads to reduced need for most medications.

So a Plant-ri-cian is a physician or clinician empowered with knowledge of the benefits of whole food plant-based nutrition.

I can most strongly recommend the Plant Based Nutrition eCornell University Course to every doctor or therapist who wants to give good guidance on diet to their patients. For me it was the best value educational training I have ever had, based on the study time involved and the benefits derived for me as a practitioner and for all my patients.

Chapter 18

Diabetes Management and Treatment: Where's the Problem Carbs or Fats?

A simple experiment has been repeated many times, two matched groups, one on a high-fat diet and the other on a carb-rich diet are challenged with a heavy dose of glucose. The first group's blood sugar far exceeds the second group. It has taken seventy years to discover the cause of Type 2 diabetes.

For blood sugar to be absorbed into our cells insulin is needed. Insulin is the key that unlocks the door to let sugar in our blood enter the muscle cell. When insulin attaches to the insulin receptor, it activates an enzyme, which activates another enzyme, which activates two more enzymes, which finally activate glucose transport, which acts as a gateway for glucose to enter the cell. So, insulin is the key that unlocks the door into our muscle cells.

What if there was no insulin, though? Well, blood sugar would be stuck out in the bloodstream, banging on the door to our muscles, and not able to get inside. And so, with nowhere to go, sugar levels would rise and rise. That's what happens in Type 1 diabetes; the cells in the pancreas that make insulin get destroyed, and without insulin, sugar in the blood can't get out of the blood into the muscles, and blood sugar rises.

But, there's a second way we could end up with high blood sugar. What if there's enough insulin, but the insulin doesn't work? The key is there, but something's gummed up the lock. This is called insulin resistance. Our muscle cells become resistant to the effect of insulin.

What's gumming up the door locks on our muscle cells, preventing insulin from letting sugar in? Fat. What is called intramyocellular lipid, or fat inside our muscle cells. Fat in the bloodstream can build up inside the muscle cells, create toxic fatty breakdown products and free radicals that can block the signalling pathway process. So, no matter how much insulin we have out in our blood, it's not able to open the glucose gates, and blood sugar levels build up in the blood.

This mechanism, by which fat (specifically saturated fat) induces insulin resistance, wasn't known until fancy MRI techniques were developed to see what was happening inside people's muscles as fat was infused into their bloodstream, and that's how scientists found that elevation of fat levels in the blood "causes insulin resistance by inhibition of glucose transport" into the muscles.

One hit of fat can start causing insulin resistance, inhibiting glucose uptake within just three hours. Then, you can do the opposite experiment. Lower the level of fat in people's blood, and the insulin resistance comes right down. Clear the fat out of the blood, and you can clear the sugar out of the blood. So, that explains this finding. On the high-fat diet, the ketogenic diet, insulin doesn't work as well. Our bodies are insulin-resistant.

But, as the amount of fat in our diet gets lower and lower, insulin works better and better. This is a clear demonstration that the sugar tolerance of even healthy individuals can be "impaired by administering a low-carb, high-fat diet." But, we can decrease insulin resistance—the cause of pre-diabetes, the cause of Type 2 diabetes - by decreasing saturated fat intake.[18]

The recommended intake of fat is about 20g. per day. On a Whole Plant Diet a person needing 2500 calories would get 20g of fat but on an animal based diet (meat, fish, dairy) they would be getting nearer 180g of fat, many times more than the required, healthy level! This is where the insulin effectiveness fails.

Most diabetics can manage on dramatically far less insulin, or often none, when they shift their diet from high animal/dairy/refined carbs towards a whole plant diet of fruit, vegetables, whole grains, pulses and some nuts and seeds. Watch Dr Neal Barnard explain this in more detail on a video.[22]

Chapter 19

A New "Wonder Drug" or Just Food For Thought?

Here's a question for you to answer: "If there was a new drug you could take that promised:
you would live a longer healthy active life,
look and feel younger,
have more energy,
lose weight, if overweight,
lower your blood cholesterol,
prevent or even reverse heart disease,
lower your risk of prostate, breast or other cancers,
preserve your eyesight in your later years,
prevent or reverse diabetes,
avoid surgery in many instances,
vastly decrease the need for other pharmaceutical drugs,
keep your bones strong,
avoid impotence,
avoid a stroke,
prevent kidney stones,
alleviate constipation,
lower your blood pressure,
avoid Alzheimer's Disease,
beat arthritis, and much more.....
and that this new drug would not cost you a fortune, but would save you thousands of pounds over your lifetime, and that there would be no adverse side-effects, so common for most drugs,
would you be willing to take part in a research trial of this drug?

Yes or No?

Well, unfortunately there is no such drug on the market. If there was, it would be an instant best seller and make the drug company a fortune. There are thousands of drugs that can help one or two of these conditions but at a cost of the usual undesirable side-effects.

However, there's no need to be disappointed because all the evidence of over forty years of research and the study of the lifestyles of hundreds of millions of people across the world suggests we may not need a drug at all, but we simply need to change the food we eat along with a few other simple lifestyle changes to achieve most of these same results.

Yes, it would seem that with the right diet we could all achieve most of the benefits outlined at the start, barring any major environmental or other unavoidable health risks.

What then would be the "right diet"?

That diet is based on eating only "Whole Plant Foods", not requiring numerous food supplements, enzyme pills, special herbals, calorie counting, extra shopping costs or expensive professional consultations.

The evidence for this is very strong. An excellent introduction to this subject by one of the key founders of the Whole Plant Diet, Dr T Colin Campbell is to be found on this video.[23] Additional hard evidence is to be found in a remarkable book by Dr Michael Greger entitled "How Not To Die" [29]

Chapter 20

You Can Prevent and Reverse Heart Disease!

The most common cause of death in the West is a heart attack; in America over 40% of deaths are caused by a heart attack, more than any other disease or injury including cancer. In the UK over 26% of deaths are from heart disease, around 160,000 every year, the second major cause of premature death after cancer. Every 24 hours some 3000 Americans will have a heart attack, roughly the same number that died in the September 11th terrorist attack in 2001. Heart disease used to be a condition of old age, today it is increasingly found among young people and even children. A study of 300 male soldiers in their 20's who had died in action had their hearts examined and it was found that 77% had evidence of gross heart disease.

What is heart disease? Perhaps the most common cause of sudden death is a heart attack or myocardial infarction. The vast majority of these sad events can be avoided with two simple lifestyle changes; improved breathing and the right diet.

Myocardial infarction, otherwise known as a heart attack, occurs when blood flow bringing oxygen to the heart is severely reduced or totally blocked. It results in oxygen starvation and damages part of the heart.

These attacks often occur during or following physical exercise or emotional stress, both these activities increase breathing volume but when breathing volume is greater than the body's needs: carbon dioxide is lost resulting in reduced blood flow and reduced oxidation of the heart muscle. In a paper entitled Hyperventilation Myocardial Infarction by Jovanovski in 1988, he wrote, in addition to causing peripheral cerebral vasoconstriction, hyperventilation has also been shown to cause diminished coronary blood flow and oxygen delivered to the heart.

Cardiac arrest is different from a heart attack, the principal cause of a cardiac arrest is electrical signals that control timing and organization of the heartbeat becoming completely chaotic and when signals degenerate into total chaos the heart suddenly stops beating, cutting off normal circulation to the body. While the causes of cardiac arrest are numerous, by far the most was common in adults, is ischemic cardiovascular disease. Breathing in excess of normal metabolic requirements causes a loss of carbon dioxide from the blood leading to hypocapnia with reduced oxygenation of the heart and disturbed cardiac rhythm.

Most heart disease is associated with a narrowing of arteries supplying blood to the heart due to the slow production of fatty plaque. At some point this hidden problem goes critical and forms a blockage on a vital artery causing the heart attack. We now know what are the main causes of a heart attack.

Risk factors include the following:

High cholesterol level in the blood. (Statins are not the best or only answer),

Stress,

High blood pressure,

Smoking,

Lack of physical exercise,

Overweight or obesity.

Of all these factors the most reliable predictor of heart disease is elevated cholesterol levels in the blood. Elevated blood cholesterol is closely linked to consumption of animal based foods.

Smoking and high blood pressure are also related to stress in people's lives and as stress has a profound effect on our health this should always be addressed and managed. Breath training is a proven way of managing the adverse effects of stress.

Thousands of scientific papers been published concerning the dietary impact on heart disease and perhaps the major relationship that is almost universally agreed as a predictor of heart disease is the cholesterol levels in the body.

An eminent American surgeon Dr Esselstyne was concerned about the poor treatment of cancer and heart disease as there was little effort at prevention but most research work focused on treatment. He began a major study looking at the effect of reducing fat and cholesterol intake of his patients. This meant almost eliminating all animal products, refined oils, dairy produce, fish and fowl. Of the group of patients that are adhered to this dietary change, the average cholesterol levels fell from 246 mg/dL at the start to below 132 mg/dL. Over the 11 years of the study only one participant had a coronary event and this was a patient who had strayed from the diet for two years; after this incident he returned to the Whole Plant-based diet and ceased to have any symptoms or further events.[19]

If you want to protect yourself from this common killer, start by shifting your diet towards a Whole Plant Based diet, eating less meat, fish, processed foods, dairy food, oils and fats and refined foods like white flour or white sugar.

You can check your progress with the "4LeafSurvey" screening test either on my website or directly online at "4leafsurvey.com" and if your score is not a plus score you should help yourself to better health and reduced risk of most chronic diseases by starting to make changes in your diet. A healthy score is over +30.

Chapter 21

Pigs Bite Back or "Pig Karma"!

There's no doubt that bacon is one of our favourite foods, the English breakfast is traditionally based on bacon and eggs with various other foods added, the bacon roll is a hot favourite as a quick snack and sausages wrapped in bacon are regarded as a real treat. Bacon has decidedly become part of our national diet, but to meet this demand from fifty plus million consumers, the farming of pigs has had to go big.

Now pig farming is no longer an idyllic country scene where the few pigs were kept in their comfy sty, fed the best food and looked after with the greatest care and attention till eventually the day comes for their demise, when even this necessary end is achieved with a respect and concern for the animal. Today pig farming is on an industrial scale and the animals are regarded more like objects to be processed efficiently and their final end must come as an immense relief to them, since they *are* sentient animals that can experience enjoyment or pain like any other mammal.

If pork in all its forms were an essential part of human diets, or if it conveyed some particular health benefit that no other food could give us or if we could not find sufficient nutrition from alternatives to pig or for that matter any animal foods, then there would be a sound case for the continued expanding production and consumption of this animal. But the reality is that none of those conditions apply; there are no nutritional benefits of pork over other foods, it is not an essential part of any person's diet and it is a resource-costly food to be produced in quantity.

Pig farming today consumes excessive amounts of energy, land and water compared with other plant-based foods; *a pound of pork requires over 500 gallons of water, 12 Kilowatt hours of energy, over 6 pounds of animal feed. A pound of pork produces the equivalent CO2 emissions to an average car driven 15 miles and the waste from pig factory farms, approximately 3000 pound a year per animal, is a serious hazard to health and causes pollution of waterways.*

So, all in all, it would seem our love of pork comes at a high price to the environment and natural resources but, hey, what the heck, we love it!

Karma is simply defined as what goes round comes round!

Maybe there is a "pig karma" since we now are told by the World Health Organization that our favourite food is killing us!

Yes! This is from the highest scientific authority!

Bacon, ham and sausages rank alongside cigarettes as a major cause of cancer, the World Health Organization has said, placing cured and processed meats in the same category as asbestos, alcohol, arsenic and tobacco.[20] This news is bad enough but the health hazard of the high fat and cholesterol content of the meat just adds to the risk of an early demise for the pork consuming enthusiasts.

The report from the WHO's International Agency for Research on Cancer said there was enough evidence to rank processed meats as Group 1 carcinogens because of a causal link with bowel cancer.

Just like the tobacco story, that took fifty years for the medical profession and governments to accept the evidence that "smoking kills", there will be many years of debate and counter argument from the meat industry before our attitudes to eating meat and especially processed meats changes our dietary habits.

There is a term "precautionary measures" that ought to apply here. It we know something is likely to be a threat to our lives and health why not avoid it if at all possible?

Smokers didn't need to wait fifty years to be told by the government that smoking is bad for your health printed on the packet, they could have chosen to play safe and quit when the first serious concerns were published. Likewise, since we do not need meat in our diet for optimal health, despite what the meat industry and their paid experts will tell you, why not quit it now and play safe?

That would be a precautionary measure that makes good sense for your long-term health and the planet's sustainability for our children and grandchildren as well.

Beware of Pig Karma!

Chapter 22

The Pisa Pelvic Exercise

Here's a tip you might like to use to keep your pelvis in good alignment with just thirty seconds of exercise each day.

After thirty-five years in practice it struck me that I didn't know why the vast majority of my patients presented with the same structural problem, a pelvis with a torsion anterior right & posterior left.

It was then that I thought about Pisa's leaning tower.

If the builder had notice the first signs of leaning soon after the construction he could have advised the mayor of Pisa to send out a bunch of strong citizens every day with long poles to push it back, in which case there might never have been a tourist attraction three hundred years on!

In fact my wife and I were among the last tourists to be allowed to go up the tower because it had become too dangerous. The good news is that a decade or two ago, engineers started to fix it, not with poles but with giant hydraulic rams! (I have recently been told the builder actually built it with a tilt, to make it unique! He didn't reckon with gravity though!)

So maybe it was gravity that was causing this torsion?

But why was it always the same way?

Was it because of the asymmetry of our heavy liver compared with lighter organs on the left? I thought I had the reason, but then couldn't understand why most people in under-developed countries had good posture and not much work for an osteopath, yet they had the same internal organs! There must be a better reason!

I had to find another explanation.

We humans are upright, and most would agree that is what is special about humans, we are "homo erectus" but maybe not so "homo sapiens"!

There are two forces acting on us all that achieve this; "gravity" that is heavy, dark, grave and pulls us down, and an opposing force, shall we call it "levity", that is light, uplifting, and happy that lifts us upwards. Some would call the latter spirit or positive emotions.

If gravity is stronger than levity we are bowed down, if levity is stronger than gravity we are no longer grounded and have our heads in the clouds, but when balanced we are upright and balanced. It also struck me that in poorer countries we invariably see smiling faces despite their poverty and hard lives, these people may not be rich but they have a strong inner strength and a happy spirit.

To finish the concept let's assume gravity is an anti-clockwise torsional force and the levity a clockwise torsional force. Thus in the West the gravity being stronger than our levity (because of our stressful lifestyle of finance pressures, work pressures, environment, consumerism, etc.) we develop the anterior torsion of the pelvis, and once started, gravity takes over just like the Pisa Tower. In less developed societies levity is stronger and maintains their good posture.

Could I find anything to substantiate this hypothesis? To try to put an observation to this idea, I noticed that everyone has a clockwise hair growth, could this be due to the torsional levity force?

How could we increase our levity force in the West?

This could be a tough challenge for most practitioners but maybe teaching patients to breathe better would reduce the effects of stress and make a difference. We can't be stressed if we breathe calmly. So that maybe one way of keeping away from your physical therapist!

If patients are not interested in sorting out their stress, how else could we help? From the Pisa problem I hit on an exercise to help eliminate them developing a pelvic torsion. I decided to try to help them with a *very simple thirty second exercise*, despite the fact I find exercises are boring and rarely recommend them except in a very small percentage of patients. I guessed a thirty-second exercise couldn't be too boring!

The Thirty Second Pisa Pelvic Exercise

The exercise involves lying on your back, taking hold of your right knee and pulling with both hands towards your chest sufficient to stretch the posterior muscles and ligaments on your right side of your pelvis for just thirty seconds, then release. End of exercise!

I began by trying this on myself and have maintained a level pelvis this way for over two years now, which is good for me, as before I used to need a treatment and a pelvic adjustment every few months.

No adverse side-effects, except it might affect your osteopath's business adversely as you will be able to go longer between treatments!

Simple to do in bed or lying in your bath, so why not try it yourself and even tell your friends about it, I do?

Remember this is intentionally an *asymmetric exercise*, just on the right side.

Who knows, perhaps from an osteopathic standpoint if we can keep our pelvis from twisting mechanically; as "structure governs function", it should even help us increase our levity!!

Chapter 23

Structure Governs Function

The philosophy underpinning osteopathy and other allied physical therapies is that it matters how we are structurally aligned to function optimally. This is simply recognizing the basic laws of mechanics that apply to every structure including our human structures. "Structure governs function" rules this matter. It is also true that "Function may govern structure" in some situations.

Despite this simple, obvious rule of mechanics, there is a remarkable lack of understanding amongst the mainstream medical community that any structural problem in their patient will be significantly impairing some bodily function, internal or external. It would seem that many doctors regard the human body as being independent of the normal rules of mechanics, truly a unique mechanical structure. To accept the proposition that many internal disorders might be helped to recovery by improving the patient's body mechanics might be beyond medical credibility, despite the fact that osteopathy was founded on these relationships.

Thus, wherever there is a misalignment or impaired movement in any part of our body we must recognize that there will be some degree of malfunction in the body, in particular associated with the particular organ or tissues served by nerves and blood supply from that segment of the spine. The following notes explain some of these associations but it must be born in mind that no single part of our body works independently of the whole body and that the impact of a structural problem in one area will affect the entire working of the body at all levels. This is the therapeutic concept underpinning all physical therapy.

How any structural problem is restored to normal will depend on the therapy used, the particular conditions associated with the patient and many more related factors. Usually physical treatment will include articulation, relaxation, release work, soft tissue massage or other approaches that support this process of normalization.

It may not be an exaggeration to say that every cell in our bodies is connected in some way with every other cell. That a problem in one area of the spine will impact on the rest of the mechanics and in particular on all internal organs supplied by nerve and blood from that segment is irrefutable.

The body will heal itself given the right condition and support except in extreme states of sickness. With good posture the body remains balanced in three dimensional space. We can stand on one foot, hold a book in one hand, read, and at the same time bring a cup to our lips without even looking, and without falling over. This is a complex task requiring a constant self-correcting mechanism. When a patient suffers a laceration, a physician can only assist by cleaning the wound, and bringing the edges together. The healing occurs on its own.

There is within each of us an inherent healing functionality that constantly works to restore homeostasis and body integrity. An integral part of this self-healing process is found in the arterial circulation of our blood. When blood and lymphatics flow freely, tissues can perform their physiologic functions without impedance. With the occurrence of trauma (physical or emotional), the tissues contract, twist, and compress, the fluid flow becomes obstructed. Micro-climates of under-perfusion result and are considered to be a significant contributor to the onset of disease. Mechanical adjustment restores freedom in the tissues, normalizes fluid flow and thus inherent physiologic function (healing) follows.

Every cell in our body requires nutrition, oxygenation and protection from foreign viruses or bacteria; the arterial blood carries all these components to every cell where physically possible. Restriction of arterial blood flow, either by physical compression of the blood vessels or by the contraction of smooth muscle tissue from nerve irritation, is perhaps a prime factor in all diseases. It was the recognition of this process that led to the development of osteopathy as a total medical system over a century ago and has since been the basis of all physical therapy positive results. It is quite remarkably that in this information rich world we now live in, just how few people, including doctors, know how our physical structure is intimately connected with our health or disease state.

How can a pain in your back be connected to a heart problem?

Any nerve root irritation in the neck will alter the nerve supply to the heart, as will problems causing similar irritation in the upper back. These effects can alter the rhythm, beat rate & strength of the heart. Likewise blood supply to and from the heart will be altered with irritated nerve supply to all the blood vessels in the body.

The reverse is also true, that any dysfunction of the heart will cause a change in the tissue corresponding to the nerve supply area of the back.

The consequences of this inter-connectedness are that:

1. When there is a tender or painful area in the back we should be aware that it may also be causing a hidden adverse effect on the functioning of any internal organ.

2. That a painful or tender area in the back may be primarily due to a problem in an internal organ and

3. By examining the back we can aid diagnosis of other health problems and by treating these same areas we can help improve the functioning of the related organs.

Once this concept is understood it is easy to see how what seems like just structural treatment, however it is carried out, may have a profound effect on the patient's general health way beyond relieving "the pain in the back".

Chapter 24

"What is an Orthopath?"

The Orthopath sees health as based on restoration of all functions and physiology to normal. Structural correction is always part of their work but structural correction is not regarded as enough.

Usually nutritional screening and advice is offered to their patients but that is still not enough to help the patient to optimal health. Advice on suitable exercise and rest is routinely included in their therapy and this adds a little more to the health promotion plan. The patient's work and environmental factors are reviewed to check whether there are health issues here that need attention.

Some orthopaths will routinely check the patient's breathing to assess any deviation from normal optimal breathing. Even all this may not uncover other hidden problems that are having an adverse effect on the patient's health that may be from stress and mental attitudes. When all these factors and more are brought together we start to give a holistic approach to health care, this is the ultimate aim of an orthopath that may never be fully realized but it is the aspiration the orthopath works towards in their practice.

Does an orthopath have training in physical therapy?

Yes just like an osteopath, chiropractor or physiotherapist.

Does an orthopath have a fundamental holistic approach?

Ideally yes, but this is not necessarily so with other physical therapists.

May an orthopath use adjunctive therapies to benefit the patient?

Yes, this by definition is an essential matter. Many other physical therapists are constrained as to what they're allowed to use or, what they want to use, in addition to their physical therapy.

Is an orthopath's work primarily based on structure?

No, unlike many other physical therapists, the orthopath tries to expand their realm of therapy and support beyond just physical structure.

Is an orthopath expected to teach the patient health promotion?

Yes indeed, this is a prime part of the orthopath's work, to try to make the patient more and more self-reliant by helping themselves to better health.

So can orthopaths advise on all aspects of health promotion?

Yes indeed, that is what most orthopaths need to do and want to do. An orthopath is constantly updating and enlarging their therapy base and exchange experiences regularly with other therapists to develop a more holistic approach to health care.

Every orthopath has to meet all legal, professional and ethical requirements like any other physical therapist.

So you could say an orthopath is an "osteopath-plus", a "chiropractor-plus", a "physiotherapist-plus", a "sports therapist-plus", etc.

Chapter 25

Enjoy Optimal Health with Lifestyle Changes

Health is both simple and complex, when anyone asks me what are the factors affecting our health, I am now quite clear that the answer is; everything. Our health is affected by many things including body mechanics, our diet, our breathing, our stress, our work, our family, exercise, the environment, and much more or as I said before, everything, but we only have any substantial control over two of these factors. The good news is that by taking control over the two key factors we may improve most of the other factors. So what are these two health-promoting activities in our lives?

They are what we eat and how we breathe!

There is an old saying "We are what we eat" or I tell children "We can't make eyeballs from chocolate bars"; they usually get the idea from that.

Today it is now recognized that bad eating habits have led to an epidemic of obesity, heart disease and diabetes but the public are still not given the facts that scientific research proved over twenty or so years ago. This is in part due to the fact that the food industry would rather keep us all in the dark, also because doctors do not have any significant training in nutrition so feel unable to offer any good advice to their patients and to be fair to the doctors, often their patients don't want to hear any advice that might involve them cutting out foods they have come to love.

However the fact is that ninety percent of our major chronic diseases are diet related, in other words we can reduce or eliminate the risk of developing the major diseases like; Heart Disease, Diabetes, Strokes, Alzheimer's Disease, Cancer, MS, Arthritis and more by changing our diet. *Yes, it's that simple and that powerful, we have the power to improve our health without drugs.*

We can live for three weeks without food, three days without water but barely three minutes without air. Would this alone not make us think that breathing is the most vital activity in our lives? Over seventy five percent of us breathe badly, most of us over-breathe or suffer some degree of chronic hidden hyperventilation, or over-breathing. Dysfunctional breathing impairs oxygenation of our body, causes restrictions of blood vessels, disturbs the entire chemistry of the body and may be associated with over a hundred diseases including Asthma, Angina, Anxiety, Hypertension, ME, Poor Sleep, IBS, Allergies, Sleep Apnoea, Orthodontic Problems, Panic Attacks and many more health problems. We can all learn how to improve our breathing in just a few weeks, because bad breathing is just a bad habit and the body is ready and waiting to return this vital activity back to normal.

The really good news is that my research over the past three years, based on over two hundred patients I have screened for breathing and diet, has shown that the way we breathe affects the way we eat and the way we eat changes the way we breathe.

This means every improvement of one will automatically lead to improvements in the other.

I have a Skype Lifestyle Course[12] that teach you how to improve both your diet and your breathing and is based on two major research works; the dietary advice is based on the largest epidemiological study ever done in the field of nutrition, backed by over twenty-five years of clinical tests and used by over 12,000 doctors and health workers worldwide, often referred to as "The China Study". The breath training is based on the clinically proven system the Buteyko Method developed by the Russian doctor Professor Konstantin Buteyko. The course is based on six half hourly sessions and usually extends over three weeks. Think about it, just three weeks to learn how to enjoy optimal health!

Chapter 26

Forget Your Drugs, Sort Your Lifestyle!

Almost all chronic diseases and most acute illnesses are the product of years of many lifestyle stressors on the body.

It is wishful thinking or just ignorance to believe we can take a single pill to fix a disease.

Let me emphasize the fact that some people may be totally dependent on certain drugs, this podcast is about the rest of us.

This trust in a drug-based solution has been driven by two main forces, the commercial profit motives of the pharmaceutical companies supported by the modern reductionist medicine, and the demands from patients to be given a "pill for every ill" rather than accept the need to make changes in their lifestyle, supported by a failure of health promotion education for both doctors and the public at large.

I am not dismissing the enormous advances that have been made in drug treatment that are keeping millions alive and in a functioning state, that without their use and without any significant change in lifestyle, most would suffer far more and meet an early death.

So what are the factors that might ensure most of us would live a long, active, disease free life? What evidence is there to support this concept? Essentially, could it be true that vibrant health is the normal state of affairs and that diseases should be the remarkable exceptions? Could this be the great medical secret?

Health is based on normal structural integrity (almost ignored in modern medicine yet the fundamental tenet of osteopathy that recognizes the fact that "structure governs function"), health is founded on the optimal diet (once again a factor almost ignored by modern medicine despite the fact that 90% of chronic diseases are closely related to diet, "we are what we eat"), health is dependent on normal breathing (few doctors routinely check their patient's breathing despite the fact that almost all diseases are linked to dysfunctional breathing and that most people in the West suffer from over-breathing or chronic hidden hyperventilation), health is dependent on an adequate level of physical activity (our sedentary lifestyle and increasing reliance on wheels over legs has become a major issue), health is severely damaged by stress or our mental outlook and state (this is the most complex factor affected by our childhood, our family, work, education, emotional traumas etc., *so it would seem pretty unlikely that such a complex, multifaceted problem could be fixed with a drug or two*), health is dependent on our environment (this is for most of us outside our individual control, pollution of the air we breathe or the water and food we consume along with many other toxins found in our modern lives, there are few drugs that can eliminate or neutralize these poisons to our body), the list could be extended indefinitely for the simple reason that health is connected to, well, everything, as I expand on in my book entitled "Connection - towards a broader understanding of health in medicine."[13]

As a first small step in this direction I offer a short Skype Course " Optimal Health Lifestyle Course"[12] that addresses the three most important factors, diet, breathing and stress.

All this is not new but can be found in the growing interest in "functional medicine" that tries to find the causes of ill health, where the body's functioning has gone astray and then to address these factors. Looks very much like the above, don't you think!

Chapter 27

Water – The Foundation of Health & Life

Here again is yet another vital factor in our health that is hardly discussed in medicine.

We can survive for weeks without food but only for a few days without water, we are 75% made of water, we don't have copper wires to carry nerve signals, our nerves are fluid wires and our brain is 85% water. Imagine the consequences of serious water loss, how it would affect the entire functioning of our body and in particular our nervous system and our ability to use our brain.

Besides this, water is a unique fluid with many strange properties we are only now beginning to understand, so it is no use thinking you can replace pure water with any other fluid or beverage to meet the body's demand for water, it doesn't work.

Most of our chosen drinks like tea, coffee, beer, wine, fizzy drinks not only don't add to the body's water supply, but usually cause increased elimination of water so that we lose more than we take in.

Most would say, don't worry because we know when we need water ,we get thirsty and have a dry mouth. Sadly this is *not true*, the thirst for water and a dry mouth is felt at the extreme end of water shortage while the body has been suffering dehydration for some time.

So what does water do for our bodies?

The greatest authority on this subject is a Doctor Batmanghelidj who spent a lifetime of research studying the role of water in our lives. In his book "You're not sick, you're thirsty!"[14] he lists forty-six reasons why your body needs water every day, I will give just a few of these reasons here; there is no life without water, water is the prime solvent for all foods, water is the transport system of nutrients for the body, water clears toxins from the body, water lubricates our joints, water is the essential fluid for our cooling system - sweating, adequate water protects our blood from clotting while circulating and water is involved in all energy exchange systems in the body.

Dehydration is almost an epidemic today laying the foundation for many common ailments from asthma and allergies to stress and strokes.

It also appears that as we age we become less sensitive to the need for water and many elderly people are often severely dehydrated.

Myths in medicine about water:

A dry mouth is the only sign of dehydration.
   The truth is, a dry mouth is the very last sign of dehydration!

Water is a simple liquid.
   Water is now recognized to be a remarkable unique fluid: it is life-sustaining and health promoting.

The body can regulate water needs and intake.
   Our perception of thirst is not reliable: as we get older it fails us.

Any fluid can replace water.
   Most other fluids: Tea, Coffee, Fizzy drinks, Alcohol, etc. not only do not replace water but cause loss of water.

Some of the key Facts:

• Every 24 hours the body recycles the equivalent of 40,000 glasses of water!
• Every day the body becomes short of 6 to 10 glasses of water.
• We all need about our weight in kilos, in fluid ounces of water per day for good hydration.

Your Guide to How Much Water You Need:

Your weight in pounds	Your weight in Kilos	Ideal Water Intake fl. oz.	Ideal Water Intake litres	Glasses of water*
80 to 100	35 to 45	40 to 50	1 to 1.5	5 to 6
100 to 150	45 to 70	50 to 75	1.5 to 2	6 to 9
150 to 200	70 to 90	75 to 100	2 to 3	9 to 12
200 to 250	90 to 115	100 to125	3 to 3.5	12 to 15

*Glass of 8 fluid ounces.

Advice:
• Drink water before eating.
• Always drink water when thirsty even during a meal.
• Drink water 2 to 3 hours after a meal.
• Drink water first thing in the morning to correct for loss during sleep.
• Drink water before exercising, to prepare for sweating.
• Drink more water if constipated; 2 to 3 glasses in the morning act as a good laxative.
• So just how much water do we need to protect ourselves from the ravages of dehydration? As a general rule the same number of fluid ounces as your weight in kilos or the same number of fluid ounces as half your weight in pounds. On average about two litres per day for the average weight person. Remember tea, coffee and beer don't count but add to the need for more water.

Chapter 28

Boost Your Sports Performance & Physical Fitness

There is a vast amount of misinformation on this subject, or as President Trump would say "Fake News". In part because of commercial forces trying to sell their wares and in part due to lack of up-to-date research.

There are two main factors that you have control over that will significantly improve your physical health and performance, they are what you eat and how you breathe.

Let's start with the most sensitive subject; what you eat.

A myth that has survived a hundred years and is still going strong is that anyone trying to improve their muscle bulk and strength needs plenty of quality meat protein or at least some form of protein from milk. Well, the truth is that we can all get the same amount of protein per 100 calories from plant-based foods and we don't need meat or milk protein.

There is more to tell; milk protein is a significant health hazard as there is a link between prostate and breast cancer and the consumption of milk.[15] There is also concern that milk consumption is associated with MS and onset of Type1 Diabetes.[16] It is the perfect food for calves and is formulated to make them grow quickly and therefore contains growth-promoting hormones. The Japanese hardly knew what prostate cancer or breast cancer were until dairy produce was introduced into their diets, but Japanese emigrating to the USA soon develop these cancers as often as the Americans.

What other downsides are there to meat eating? Well meat contains very few complex nutrients such as vitamins as they have already been digested by the animal themselves, neither does meat provide any fibre, essential for both a healthy gut and to feed your gut bacteria, our micro-biomes. There are many top world athletes, iron men and body builders who manage very well on a diet free of meat and dairy. They are in good company as some of our greatest mammals do pretty well on greens and grains like the gorilla and elephant![48]

Now let's consider breathing as an aid to fitness and endurance. Once again there is a long held myth that deep breathing is beneficial for physical performance. Wrong again! Normal breathing is optimal for this. What, though, is normal breathing? From my observations of hundreds of my patients over the past fifteen years I have found only a few percent who actually breathe normally, 90% are over-breathing or suffering from some degree of Chronic Hidden Hyperventilation (CHHV), The 21st Century Epidemic.[17] This includes the sports people and apparently healthy fit people. Normal breathing is defined as always breathing through the nose and breathing around eight to ten breaths per minute and five to six litres per minute at rest. We have a culture of training from childhood to adulthood that deep breathing is good, that bigger breaths lead to better health, this is a myth that the Buteyko Method dispels once and for all.

You might like to go through the check list of adverse reactions to CHHV. Do you recognize any of the following symptoms in yourself?:

Poor stamina & endurance
Frequent muscle cramps with exercise
Low energy levels
Poor sleep quality
Dehydration or dry mouth
Recurrent infections or poor recovery
Muscle stiffness
Blocked or runny nose
Hypertension
Itching, dry skin, eczema
Craving for sweet snacks
Irritability
Mild depression

Headaches
Excessive mucous production
Allergies or hay fever , these are some but not all of the
symptoms associated with CHHV.

Check Your Own Breathing Now: Sitting quietly, keep your mouth closed, take a normal breath in through your nose, then exhale normally through the nose. Hold your nose and hold your breath until you feel you need to breathe in again, this is a maximum **comfortable** breath hold, **not as long as you can hold your breath.** Release your nose and breathe in gently through your nose. Check the number of seconds you held your breath. This is called the Control Pause (CP) How did you do?

If your CP is under 15 seconds you are breathing enough for two to four people and your body is very poorly oxygenated,

between 15 and 25 seconds that's enough for two people and you may have health problems such as asthma, hypertension, etc.

between 25 and 35 seconds you are breathing 50% more than you need and your body oxygenation is poor,

between 35 to 45 seconds that's about normal but over 45 seconds is what we need for peak oxygenation and optimal performance, ideally 45-60 seconds.

Chapter 29

The Other "Five a Day" for Children

Every parent has become increasingly aware of the need for good nutrition, good exercise and a supportive home and school environment for the health and wellbeing of their children.

The government campaigns have advised "five a day" and more sport in schools but as yet there has been no recognition of the equally vital matter of good breathing for good health.

Poor breathing has been the hidden, silent factor responsible for many childhood health problems that can lay the foundation for a wide range of modern diseases in later life. Here is another "five a day" recommendation to help your child:

1. Ensure your child is a good breather.

A simple test is the "Step exercise": Get your child to take a normal breath in then a normal breath out, then while they hold their nose see how many steps they can do, keeping their mouth closed before they need to release their nose to take another breath.

This is a simple measure of how well oxygenated their body is:

120 to 80 steps indicates excellent breathing

60 to 80 steps is very good

40 to 60 steps is good

30 to 40 steps is poor and will be impairing their heath

20 to 30 steps is very poor with almost certain adverse effect on their health

Fewer than 20 steps is a dangerously low result and efforts should be made to correct this.

Older children may be able to measure their breathing with a "Control Pause". Keeping the mouth closed, take a normal breath in then a normal breath out and see how many seconds they can hold their nose for before taking another breath in.

This exercise should be easy and stress free, it is a measure of their maximum COMFORTABLE breath hold.

45 to 60 seconds Excellent,

35 to 45 Very good,

25 to 35 Good,

20 to 25 Poor,

15 to 20 Very poor,

10 to 15 Seriously poor and

Under 10 suggests an urgent need to correct this.

Why not check your own breathing with a Control Pause as well?

2. Help teach better breathing habits.

Encourage quiet nose breathing all the time. Set a good example by improving your own breathing!

3. Watch for signs of bad breathing habits.

The signs of poor breathing include, mouth breathing, upper chest breathing, breathlessness, snoring at night.

Set a good example by being aware of your own breathing faults!

4. Encourage relaxation and quiet when stressed.

Teach quiet relaxation, breathing as gently as possible with all the body relaxed. Set a good example by learning to relax yourself!

5. Tell them why they need a nose, & how to make it work well.

The nose is for breathing & the mouth is for eating and talking. "They should breathe through their mouth as often as they eat through nose!"

Teach nose-clearing exercise: Breathe in then out through the nose, keep the mouth closed, hold the nose, gently nod the head until they need to breathe in, release the nose and take a breath in through their nose. Repeat two or three times.

Check their progress from time to time with (1) above.

If your child, whatever age, mouth breathes rather than using the nose to breathe most of the time, they could be heading for many health problems that could be easily avoided.

Children who habitually mouth-breathe will usually develop crooked, crowded teeth and fail to develop normal facial features. They will be *more at risk of developing asthma* especially if there is a family history of asthma. They will probably *suffer more throat and chest infections* than others simply because they are not using their first line of defence, their nose; nose breathing kills most airborne bugs and aids the immune system.

As they get older and continue to mouth breathe they will begin to suffer the many health problems of chronic hidden hyperventilation, *anxiety, panic attacks, hypertension, heart diseases, angina, hay fever, gastric problems, breathlessness, low energy,* and the list goes on.

Finally there is a strong connection between what we eat and how we breathe, so this additional "Five a Day" is going to help with the original "Five a Day" and any improvement in diet will lead to an improvement in breathing.

Don't take our word for this, watch and hear a number of children from a Glasgow school who have trained this way, telling their own story on a video.[28]

All this is not new, the dangers of mouth breathing were brought to the attention of the medical profession over a hundred years ago in a book published, not by a doctor, but by an artist who recognized the problem while painting the indigenous population of the Americas. "Shut your mouth and save your life"[3].

Visit my website : <www.TotalHealthMatters.co.uk> to learn more about the health connections with our breathing, and if you want to understand the orthodontic relationship watch the video by Dr John Flutter, an Australian dentist.[2]

Chapter 30

Your Challenge to Save The Planet & Your Health

If everyone on the internet joined forces to ensure a healthy planet for future generations it could solve our survival problem that no government dare tackle.

All you need to do is to commit to a changing lifestyle over twelve months – a twelve-step programme, just a month at a time. Check the references for tips!

January: Try to avoid buying anything made of or packed in plastic.[34]

February: Buy local produce wherever possible.[35]

March: Cut back on your consumption of "stuff".[36]

April: Help plant a hundred trees.[37]

May: Support your local community any way that helps the environment.[38]

June: Cut out all meat, fish and dairy foods from your diet.[39]

July: Conserve fresh water whenever possible, it's a scarce resource.[40]

August: Make your own health your responsibility.[41]

Sept: Reduce the use of your car by walking more/using public transport.[42]

October: Turn your home thermostat down a few degrees lower.[43]

November: Try to reduce food waste to close on nil.[44]

December: Switch your car to any that halves your fuel consumption.[45]

If you want to know what the impact of the above would have on the planet's future survival as a place fit for human habitation read on!

January: Plastics are based on fossil fuel for production and produce an almost permanent environmental pollutant that will take many years to start to rectify. There are bio-degradable alternative materials produced from plants.

February: This would both stimulate farmers to shift to plant food production for humans, reduce the vast energy consumption involved in shipping food to and from warehouses and from abroad.

March: We all are encouraged to consume or buy far more material things than we really need, we need to remind ourselves that every article represents a substantial use of the earth's scarce resources.

April: With the estimated 4 billion internet users worldwide this would produce 400 billion trees, the most efficient carbon dioxide consumers would remove around 8 billion tons of CO_2 per year, that would help balance the residual fossil fuel use. Global production of CO_2 from transport is about this quantity.

May: There is a close connection between the health of the planet, the health of individuals and the health of a community. By building stronger communities we would be find mutual support in building a sustainable future for our only planet, the earth.

June: This alone would cut greenhouse gases by between 25 to 50%, stop the loss of the earth's lungs – the rainforests, reduce the need for food production by 75%, with the right distribution system ensure adequate food for everyone, improve the health of everyone, reduce the incidence of chronic diseases, reduce the pollution of coastal waters from agricultural medicine run-off and animal waste, permit the regrowth of our fauna and flora, especially endangered species, and much more. Currently estimates suggest between 50 – 75% of all grain and pulses go to feed animals!

July: Fresh water will be more valuable than oil in a few years' time. Sources of fresh water are increasingly being depleted due to pollution or over exploitation. Producing animal and dairy foods is a major reason for this.

August: If you have already shifted your diet to a whole plant diet and are getting more exercise walking you are well on the way to meeting this challenge.

September: This will go towards the August challenge; taking more responsibility for your health.

October: Just a few degrees drop will save you over a hundred pounds a year and reduce the consumption of energy nationally.

November: Currently we waste over a third of our food. That would feed most of the undernourished or starving throughout the world.

December: Today there is no technological reason why we cannot produce cars that will give over 100 miles per gallon; even this would make a significant reduction in CO_2 emissions. This needs to be the most important criteria for car purchase, not its sporty performance.

Afterword

This book is the accumulation of half a century of my thoughts on the subject of my passion, "health". Over that time I have studied various therapies that help support or foster health, I have read many books by much wiser authors than me on the subject of health and you can see some of them in my reading list, I have had the privilege to meet thousands of people who have come to me for some form of health care and support, and learnt from them as much as from other sources.

What I have concluded is that health has been a most neglected subject for research, has been poorly funded for the provision of any health-promoting services and is perhaps the most ill-used word in our language. E.g. We speak of out NHS, the National Health Service yet it is primarily a very good sickness service and seriously lacks health promotion, education and support. I hope that this small compilation of articles will lead the reader to re-assess the nature of health and healthcare and will help encourage others to re-assess for themselves what health means for them, whether they are doctors, health workers or members of the wider public.

If we ignore the fundamentals of health we shall continue on a downward spiral of increasing sickness and disease regardless of how much money is invested or how many more wonder drugs are discovered. All this will generate great business opportunities for the "sickness industry" but will strain our economic system beyond breaking point. Today, we have in this country and in other western nations in particular, an unsustainable healthcare system.

All the doom-laden signs are out there for all to see, but with the right thinking, actions and a concerted effort from the public, the government and the medical profession we shall be able to meet this enormous challenge.

We have to grasp this simple truth, that we must all be more responsible for our own health and our family's health. Our leaders must support us in this, by shifting resources from disease management & treatment towards health promotion and education. This will be a politically very difficult paradigm shift that will require a consensus across all political parties. It has taken almost a century to arrive at this situation and it may take a few decades to rebalance the system so that we may enjoy the best of high tech medicine as well as low tech health promotion.

References:

1. Visit www.orthopathy.org
2. Dr John Flutter video: https://youtu.be/tVjMgVClyPA
3. Download pdf. Here:
   http://www.members.westnet.com.au/pkolb/indians.pdf
4. Watch the video by Dr John Flutter. Here:
   https://youtu.be/tVjMgVClyPA
5. "The Starch Solution" by Dr John A. McDougall ISBN 978-1-62336-027-6
6. The Skype Training Course, for details visit:
   www.totalhealthmatters.co.uk
7. The Peckham Experiment, more information and history: https://thephf.org/
8. Dr Glenn Doman watch this video:
   https://youtu.be/XDdWiY6xje0
9. "The China Study" by Dr T Colin Campbell ISBN 1076-5-510-319496
10. eCornell University Course " Plant Based Nutrition Certificate Course"
    https://www.ecornell.com/certificates/nutrition/plant-based-nutrition/
11. 4LeafSurvey The online check is available here:
    www.4leafsurvey.com
12. Skype Lifestyle Course, for more details vis:
    www.totalhealthmatters.co.uk
13. "Connection" by Michael Lingard ISBN 978-1-326-94022-5
14. "You're not sick, you're thirsty!" by Dr Batmanghelidj ISBN 0-446-69074-0
15. Breast Cancer & Diet, check out :
    https://nutritionfacts.org/topics/breast-cancer/
16. Type1Diabetes,
    https://www.youtube.com/watch?v=uDQYxdJblio

17. CHHV, The 21st Century Epidemic.
https://www.youtube.com/watch?v=prYHSXsPoX0
18. Diabetes and Nutrition:
https://www.youtube.com/watch?v=ktQzM2IA-qU&t=14s
19. Heart Disease:
https://www.youtube.com/watch?v=AYTf0z_zVs0&feature=youtu.be
20. Cancer & meat:
https://www.thelancet.com/pdfs/journals/lanonc/PIIS1470-2045%2815%2900444-1.pdf
21. Health is Based on Family & Community
https://yourhealthinyourhands.simplecast.com/episodes/health-is-based-on-family-and-community
22. Dr Neal Barnard Diabetes:
https://youtu.be/ktQzM2IA-qU
23. Dr T. Colin Campbell: https://youtu.be/XEuRMm-a6mo
24. Understanding Asthma: A different Point of View:
https://youtu.be/N-9FIEsK-3k
25. Escape From Asthma podcast:
https://dashboard.simplecast.com/episodes/a0401f9c-80fc-44d4-b96b-d3995fa02f56
26. Wikipedia:
https://en.wikipedia.org/wiki/Physical_fitness
27. TotalHealthMatters! Lifestyle Course:
www.totalhealthmatters.co.uk/Lifestyle%20Training%20Course.pdf
28. Buteyko & Glasgow Children :
https://www.youtube.com/watch?v=WciSqYk0TDE&feature=youtu.be
29. How Not to Die by Dr Michael Greger. ISBN 978-1509-852-505

30. Whole Plant Nutrition: https://nutritionstudies.org/whole-food-plant-based-diet-guide/
31. Global warming: http://www.thefoodconnection.org.uk/Global%20warming.html
32. Better Breathing Means Better Health: https://better-breathing-means-better-health.simplecast.com/episodes/your-breathing-the-most-neglected-facto
33. The Brain's Way of Healing Norman Doidge ISBN 978-1-846-14424-0
34. The Plastic Problem: https://4ocean.com/
35. Local Produce: https://foodrevolution.org/blog/why-buy-local-food/
36. Overconsumption: https://www.ecowatch.com/overconsumption-fast-fashion-2399956999.html
37. Tree Planting: https://www.carbonfootprint.com/plantingtrees.html
38. Local Community: https://www.groundwork.org.uk/local-environment-everyones-responsibility
39. Why I Don't Eat Much Meat, Fish & Dairy: https://www.youtube.com/watch?v=wRmo6VNjLCk&feature=youtu.be
40. Conserve Fresh Water: https://friendsoftheearth.uk/natural-resources/13-best-ways-save-water-stop-climate-breakdown
41. Make Your Health Your Responsibility: http://www.totalhealthmatters.co.uk/
42. Walking For Health: http://www.totalhealthmatters.co.uk/
43. Thermostat: https://lifehacker.com/five-reasons-you-should-lower-your-thermostat-backed-b-1525010287

44. Food Wastage: https://olioex.com/food-waste/food-waste-facts/
45. Car Economy: https://www.confused.com/buy-a-car/best-worst/most-economical-cars
46. Kaqun Water: https://www.kaqun.co.uk/
47. Glue Ear Treatment: http://www.sinusearinfections.com/
48. The Game Changers: Watch YouTube Trailer https://youtu.be/iSpglxHTJVM

Reading list: Some of the books that have inspired me.

Title	Author	ISBN
4LeafGuide	Graff & Hicks	978-1-5076-1341-2
A Model of Health	Roy Gillett	0-947-878-25-4
A New Renaissance	David Loriman	978-086315-759-2
Ageless Body Timeless Mind	Deepak Chopra	0-7126-5673-1
Alexander Technique	Glynn Macdonald	0-00-713385-5
Anthroposophical Medicine	Dr. Michael Evans	0-7225-2771-3
Awareness through movement	Moshe Feldenkrais	0-06-062344-6
Bees	Rudolf Steiner	0-88010-457-0
Being Me and Also Us	Alison Stallibrass	0-7073-0599-3
Biology of Belief	Bruce Lipton	0-9759914-7-7
Breaking the Food Seduction	Neal Barnard	987-0312-31494-1
Catching The Light	Arthur Zajonc	0-19-509575-8
Chaos	James Gleick	978-0-718-18565-5
Climate change	Prince of Wales	978-0-718-18565-5
Coming Home to Self	Nancy Verrier	978-1-905664-81-8
Connection	M Lingard	978-1-326-94022-5
Conversations with God	N D Walsch	0-340-76544-5
Cracked	James Davies	978-1845831556-3
Disease Proof	David L Katz	978-1-59463-124-5
Disease Proof Your Child	Dr Joel Furman	978-0-312-33808-4
Dr Neal Barnard's Reversing Diabetes	Dr Neal Barnard	978-1-59486-810-8
Dr Neal Barnard's Cookbook Diabetes	Dr Neal Barnard	978-1-62336-929-3
Emotional Intelligence	Daniel Goleman	0-7475-2830-6
Eradicate Asthma Now - With Water	Dr. Betmanjeldi	1-903571-354-9
Fit Baby, Smart Baby, Your Baby	Dr Glenn Doman	978-0-7570-0376-9
Foodwise	Wendy E Cook	978-1-905570-23-2

Forks over Knives The Cookbook	Del Sroufe	61519-187-1-061-4
Forks over Knives	Gene Stone	978-1-6119-045-4
Frogs into Princes (NLP)	Richard Bandler	0-911226-18-4
Full Planet, Empty Plates	Lester R Brown	978-0-393-34415-8
Fundamentals of Therapy	Rudolph Steiner	0-000-000-000
Fuzzy & Neuro fuzzy systems	Teodorescu	0-8493-9806-1
Gaia	James Lovelock	978-0-19-878488-3
Gut Bliss	Robynne Chutkan	978-1-58333-551-2
Health & a Day	Lord Horder	0-000-000-0
Healthy Eating Healthy World	J. Morris Hicks	978-193666104-6
How Not to Die	Dr Michael Greger	978-1-250-06611-4
How Not To Die Cook Book	Michael Greger	788-1-50984-433-3
How the leopard .... (Peckham)	Brian Goodwin	0-75380-171-x
How to give your baby....	Glen Doman	0-7570-0376-9-182-6
How to Know Higher Worlds	Rudolph Steiner	0-88010-372-1
How to Multiply Baby's Intelligence	Dr Glenn Domam	978-1-910496-29
How to Teach Your Baby Math	Dr Glen Doman	0-7570-0189-0
Hyperventilation Syndrome	Dinah Bradley	1-85626-295-2
I Ching	Sam Reifler	0-553-13677-
Life Lessons	E Kubler-Ross	0-7432-0811-0
Love, Medicine and Miracles	Bernie Siegel	0-7126-7046-7
Made for Goodness	Desmond Tutu	978-006-170660-8
Medicine - Mythology & Spirituality	Dr R Twentyman	1-85584-182-7
Milk The Deadly Poison	Robert Cohen	0-9659196-0-9
Morphic Resonance	Rupert Sheldrake	978-159477317-4
Natural Grace	Rupert Sheldrake	0-385-48356-2
Nature Cure	Henry Lindlahr	1-59224-070-4
Natures Alchemist	Anna Parkinson	978-0-7112-2767-5
Plant Based Cookbook	T Sebben-Krupka	978-0-2412-3003-9
Plant Powered Families	Dreena Burton	978-1-941631-04-1

Plant Powered for Life	Sharon Palmer	978-1-61519-187-1
Plant Strong	Rip Esselstyn	978-1-4555-0935-5
Prevent & Reverse Heart Disease	Dr C B Esselstyn	978-1-58333-300-6
Program for Reversing Diabetes	Dr Neal Barnard	978-1-63565-127-0
Recognising Health	Kenneth Barlow	0-9513171-0-5
Reversing Heart Disease	Dr Dean Cornish's	0-394-57565-2
Science & Spiritual Practices	Rupert Sheldrake	978-1-473-63007-9
Science Set free	Rupert Sheldrake	978-0-7704-3672-8
Science Synthesis & Sanity	G Scott Williamson	0-7073-0259-5
Shut Your Mouth & Save Your Life	George Catlin	978-1332-003808
Stop Feeding Your Cancer	Dr John Kelly	978-0-9927798-6-3
Tai Chi	Paul Brecher	0-00-710339-5
The Campbell Plan	Dr T Campbell	978-1-62336-410-8
The Carbon Dioxide Syndrome	Jennifer Stark	0-473-09610-2
The Caring Physician	F W Peabody	9780-6740-9738-4
The Cheese Trap	Dr N D Barnard	978-1-4555-9468-9
The China Study	Dr T C Campbell	1076-5-510-319496
The China Study All Star Recipes	Leanne Campbell	978-193952997-8
The China Study Family Cookbook	Del Sroufe	978-1-944648-11-4
The China Study Quick Cookbook	Del Sroufe	978-194036781-3
The Connectivity Hypothesis	Ernst Laslow	0-7914-5786-9
The Dancing Wu Li Masters	Gary Zukav	0-7126-3817-2
The End of Dieting	Dr Joel Furman	978-0-06-224933-3
The Engine 2 Diet	Rip Esselstyn	978-0-446-506694
The Field of Form	Lawrence Edwards	0-903540-50-9
The Gentle Art of Blessing	Pierre Pradevand	2-88353-137-4
The Healing Power of Water	Masaru Emoto	978-1-4019-0877-5
The Healthiest Diet on the Planet	Dr John McDougall	978-0-06-2426765
The Hidden Messages in Water	Masaro Emoto	1-58270-114-8
The Living Energy Universe	Gary Schwartz	1-57174-455-x

The Low Carb Fraud	T Colin Campbell	978-194035309-7
The Microbiome Diet	Raphael Kellman	978-0-7382-1811-3
The Microbiome Solution	Robynne Cutkan	978-1-58333-576-5
The New McDougall Cookbook	Dr John McDougall	0-525-93610-6
The Patient Not The Cure	Dr M G Blackie	0-356-08312-8
The Pleasure Trap	Douglas J Lisle	978-1-57067-197-5
The Reverse Diabetes Diet	Dr Neal Barnard	978-1-905744-57-2
The Salt Fix	Dr J Dinicolantonio	978-0-349-41738-7
The Starch Solution	Dr J McDougall	978-1-60961-393-8
The Systems View of Life	F Capra & P Luisi	978-1-107-01136-6
Vaccination	Viera Scheibener	0-646-15124-x
What is Wisdom	Cyril Upton	978-0900001024
Whole	T Colin Campbell	978-193785624-3
Why Zebras Don't Get Ulcers	Sapolsky	0-7167-3210-6
Wine is the Best Medicine	Dr. Maury	0-285-62250-1
You Are The Universe	Deepak Chopra	978-1-84604-530-1
You're Not Sick You're Thirsty!	Dr. Batmanghelidj	0-446-69074-0
Your Body's Many Cries for Water	F Batmanghelidj	0-9702458-8-2